T0210643

European Studies in Philosophy of Science

Volume 2

Series editors
Dennis Dieks, Institute for History & Foundations of Science, Utrecht University, The Netherlands
Giovanni Gallavotti, Dipto. Fisica, Università di Roma, La Sapienza, Italy
Wenceslao J. Gonzalez, Ferrol, Spain

Editorial Board
Daniel Andler, University of Paris-Sorbonne, France
Theodore Arabatzis, University of Athens, Greece
Diderik Batens, Ghent University, Belgium
Michael Esfeld, University of Lausanne, Switzerland
Jan Faye, University of Copenhagen, Denmark
Olav Gjelsvik, University of Oslo, Norway
Stephan Hartmann, University of Munich, Germany
Gurol Irzik, Sabancı University, Turkey
Ladislav Kvasz, Charles University, Czech Republic
Adrian Miroiu, National School of Political Science and Public Administration, Romania
Elizabeth Nemeth, University of Vienna, Austria
Ilkka Niiniluoto, University of Helsinki, Finland
Samir Okasha, University of Bristol, UK
Katarzyna Paprzycka, University of Warsaw, Poland
Tomasz Placek, Jagiellonian University, Poland
Demetris Portides, University of Cyprus, Cyprus
Wlodek Rabinowicz, Lund University, Sweden
Miklos Redei, London School of Economics, UK
Friedrich Stadler, University of Vienna, Austria
Gereon Wolters, University of Konstanz, Germany

This new series results from the synergy of EPSA - European Philosophy of Science Association - and PSE - Philosophy of Science in a European Perspective: ESF Networking Programme (2008–2013). It continues the aims of the Springer series "The Philosophy of Science in a European Perspective" and is meant to give a new impetus to European research in the philosophy of science. The main purpose of the series is to provide a publication platform to young researchers working in Europe, who will thus be encouraged to publish in English and make their work internationally known and available. In addition, the series will host the EPSA conference proceedings, selected papers coming from workshops, edited volumes on specific issues in the philosophy of science, monographs and outstanding Ph.D. dissertations. There will be a special emphasis on philosophy of science originating from Europe. In all cases there will be a commitment to high standards of quality. The Editors will be assisted by an Editorial Board of renowned scholars, who will advise on the selection of manuscripts to be considered for publication.

More information about this series at http://www.springer.com/series/13909

David Ludwig

A Pluralist Theory
of the Mind

 Springer

David Ludwig
Vrije Universiteit Amsterdam
Amsterdam, The Netherlands

European Studies in Philosophy of Science
ISBN 978-3-319-37077-4   ISBN 978-3-319-22738-2 (eBook)
DOI 10.1007/978-3-319-22738-2

Springer Cham Heidelberg New York Dordrecht London
© Springer International Publishing Switzerland 2015
Softcover re-print of the Hardcover 1st edition 2015
This work is subject to copyright. All rights are reserved by the Publisher, whether the whole or part of
the material is concerned, specifically the rights of translation, reprinting, reuse of illustrations, recitation,
broadcasting, reproduction on microfilms or in any other physical way, and transmission or information
storage and retrieval, electronic adaptation, computer software, or by similar or dissimilar methodology
now known or hereafter developed.
The use of general descriptive names, registered names, trademarks, service marks, etc. in this publication
does not imply, even in the absence of a specific statement, that such names are exempt from the relevant
protective laws and regulations and therefore free for general use.
The publisher, the authors and the editors are safe to assume that the advice and information in this book
are believed to be true and accurate at the date of publication. Neither the publisher nor the authors or the
editors give a warranty, express or implied, with respect to the material contained herein or for any errors
or omissions that may have been made.

Printed on acid-free paper

Springer International Publishing AG Switzerland is part of Springer Science+Business Media
(www.springer.com)

For Megan

In this sense, however, every sensible and philosophically honest world view must be pluralistic. For the universe is variegated and manifold, a fabric woven of many qualities no two of which are exactly alike. A formal metaphysical monism, with its principle that all being is in truth one, does not give an adequate account; it must be supplemented with some sort of pluralistic principle.

Moritz Schlick – General Theory
of Knowledge, 1918/1974, p. 333

Acknowledgements

The issues raised in this book have been with me since my time as an undergraduate student at Free University Berlin. When I decided to study philosophy in 2002, my intellectual landscape was dominated by a handful of critical theorists in the tradition of the Frankfurt School. My first contact with contemporary academic philosophy came as a shock that impressed me so deeply that I quickly changed my enrollment from philosophy and ancient history to philosophy and cognitive science. As far as I remember, this decision was made during Holm Tetens' lecture on "Computers, Brains, and other Machines" and one of his melancholic pleas for a naturalist theory of the mind. Holm's uncompromising insistence on intellectual honesty left us students with the pessimistic suggestion that naturalism was a tragic truth, but a truth nonetheless. The anti-naturalists' obsession with qualia and other supposedly irreducible entities was well motivated but unfortunately not well justified.

Challenged by this pessimistic naturalism, I decided that my philosophical interests require more than anecdotal knowledge about the natural sciences. The cognitive science program at the University of Potsdam offered a great opportunity to learn about a large range of fascinating disciplines from excellent teachers like Johannes Haack and Reinhold Kliegl. Now focusing on debates about mind and cognition, my new subjects suggested a neat division of labor. As a student of cognitive science, I would learn about empirical problems of human memory, perception, problem solving, and so on. As a student of philosophy, I would be concerned with the remaining metaphysical problems such as consciousness, intentionality, self-knowledge, or free will.

However, the problems with this division of labor soon started to appear. While I did my best to follow lectures from neurolinguistics to artificial intelligence, the convenient philosophical distinction between "easy" empirical problems and "hard" metaphysical problems made less and less sense. On the one hand, the "easy" empirical problems weren't so easy after all and the reductive explanations from my *Introduction to Philosophy of Mind* courses nowhere to be found. On the other hand, there appeared to be a wealth of fascinating empirical research on consciousness, intentionality, and other entities that were claimed to fall in the philosophers' realm of competence. Of course, I was soon introduced to a number of philosophical strategies of explaining this discomfort away. Models of reductive explanation in philosophy of mind were not supposed to provide a realistic account of explanations in scientific practice. Furthermore, the "hard problem of consciousness" did not imply that scientists have nothing to say about consciousness. Instead, one just had to separate the empirically unproblematic issues such as access consciousness or neural correlates from the philosophically troubling issue of phenomenal consciousness.

My discomfort about the relation between philosophy of mind and the reality of scientific practice might have faded without three lucky coincidences during my time as student in Berlin. First, I started to work as a student assistant for Peter Bieri who had been highly influential in establishing analytic philosophy of mind in German philosophy but had become disenchanted with the entire discipline. I do not know whether Peter Bieri would agree with the core claims of this book, but our discussions about Wittgenstein, Goodman, and the state of philosophy of mind were the single most important event in the formation of the thoughts of this book. The second lucky coincidence was a seminar on mereology with Olaf Müller at the same time as I was exposed to Peter Bieri's staunch criticism of the analytic mainstream that he so successfully co-established in Germany. Olaf Müller's broadly Putnam-inspired criticism of current developments in ontology provided me with a much-needed theoretical framework to formulate my ideas. My term paper on "mereological physicalism" was a first attempt to articulate my discomfort with current philosophy of mind and to sketch a broadly pluralist alternative.

The third lucky coincidence was that I started working as a student assistant for Tania Munz at the *Max-Planck Institute for the History of Science*. I rarely talked with history of science colleagues about philosophy of mind and I rarely talk about history of science thorough this book. Still, this book would not have been possible without a perspective on science that I do not owe to philosophers but to historians. I had learned to approach philosophy of mind by starting with sweeping metaphysical doctrines (mostly physicalism and dualism in their countless varieties) and by looking at the cognitive sciences for support. It took history of science to invert this perspective. Instead of asking how science can validate a presupposed metaphysical picture, I became increasingly convinced that metaphysics needs to adapt to the reality of scientific practice. Suddenly, my point of departure was not physicalism or dualism anymore but the overwhelming diversity of ontologies and methods in contemporary science. This book is an attempt to make sense of this diversity. While I do not mean to deny the value of reduction or ontological unification in science, I present a picture that differs quite dramatically from the metaphysical accounts that have dominated philosophy of mind since the second half of the twentieth century.

Of course, I would like to think that this book is more than the sum of lucky coincidences during my philosophical education. Still, many issues have been with me for so long that I could easily include half of my academic teachers and fellow students in the acknowledgements. As I would surely forget most of them, I just want to single out one student who has become a wonderful colleague over the years: David Löwenstein has not only always given me valuable feedback but has patiently listened to me rambling about pluralism and philosophy of mind for nearly a decade.

The first step towards this book was my Magister thesis *Philosophie des Geistes ohne Ontologie* that was supervised by Holm Tetens and Georg Bertram at Free University Berlin (2010). The first version of the actual book manuscript was prepared as my doctoral thesis at Humboldt University Berlin under supervision of both Olaf Müller and Holm Tetens (2012). Olaf Müller and Holm Tetens have been wonderful teachers who always found the right balance between giving me the necessary freedom to develop my own ideas and challenging me with constructive criticism.

The final stage of preparing the book manuscript was Columbia University where I was a postdoctoral visiting scholar under supervision of Philip Kitcher from 2012 to 2014. Although the main focus of my work at Columbia was a new project on scientific ontologies, Philip's work on pragmatism and pluralism has been invaluable for clarifying basic assumptions in this book. Furthermore, I was lucky to join a great group of international scholars (especially Marcus Ohlström, Shashikala Srinivasan, and Joseph Li Vecchi) who critically questioned my arguments. Finally, Sophia Davis, Marcus Ohlström, Olivier Sartenaer, Sabine Schmidt, and an anonymous reviewer read large parts of the manuscript and challenged me to improve my arguments.

Contents

Part I   Pluralism and Scientific Practice

1   Beyond Placement Problems 3
   1.1   Varieties of Naturalism 5
   1.2   Reduction and Reductive Explanation.. 9
   1.3   Epistemological, Metaphysical, and Conceptual Pluralism 12
   1.4   The Argument in a Nutshell... 14
   References 16

2   A Historical Diagnosis 19
   2.1   Schlick's Challenge 21
   2.2   The Forgotten Mainstream of Psychophysical Parallelism 22
   2.3   The Ontological Priority of the Physical 25
   2.4   The Placement Problem in Contemporary Philosophy of Mind 27
   References 28

Part II   In Defense of Conceptual Relativity

3   Conceptual Relativity in Philosophy 33
   3.1   The Idea of a Fundamental Ontology 35
   3.2   Putnam's Case for Conceptual Relativity 39
   3.3   Understandability and the Epistemic Challenge.......................... 41
   References 44

4   Conceptual Relativity in Science 47
   4.1   Species... 48
   4.2   Cognition 56
   4.3   Intelligence 65
   4.4   What about Natural Kinds? 69
   4.5   Realism and Existential Relativity 76
   4.6   Reconsidering the Dialectical Situation 78
   References 80

5 The Demarcation Problem of Conceptual Relativity 85
     5.1   Verbal and Substantive Disputes 86
     5.2   Interpretive Charity as an Answer to Demarcation Problem 88
     5.3   Turning the Demarcation Problem Upside Down 93
     5.4   Joint Carving and Similarity 95
     5.5   From Conceptual Relativity to a Pluralist Theory of the Mind? 98
     References 99

Part III   From Conceptual Relativity to Vertical Pluralism

6 The Argument from Horizontal Pluralism 103
     6.1   Does Horizontal Pluralism Imply Vertical Pluralism? 106
     6.2   Dupré's Promiscuous Realism 109
     6.3   Bridge Principles and Notions of Reduction 112
     6.4   Horizontal Pluralism and Multiple Realization 116
     6.5   Reductive Explanation without Reduction 121
     6.6   Reductive Explanations of Elementary Learning in *Aplysia* 125
     6.7   Limits of Reductivism 128
     References 133

7 The Argument from Ontological Non-fundamentalism 137
     7.1   Notions of Ontological Priority 140
     7.2   Supervenience-Based Formulations of Ontological Priority 142
     7.3   Reductivism, Non-reductivism, and Anti-reductivism 148
     References 150

Part IV   Beyond the Mind-Body Problem

8 Consciousness 155
     8.1   Revisiting the Hard Problem of Consciousness 156
     8.2   The Common Puzzles 158
     8.3   The Uniqueness of Phenomenal Concepts 162
     8.4   Phenomenal Concepts and Physicalism 165
     References 170

9 Beyond Dualism and Physicalism 173
     9.1   But Isn't This Dualism? 173
     9.2   But Isn't This Physicalism? 175
     9.3   The Identity Objection 178
     9.4   Limits of Identity 180
     References 185

10 Mental Causation 187
     References 191

11 Epilogue Metaphysics in a Complex World 193
     11.1 Towards an Empirically Grounded Philosophy of Mind 193
     11.2 Metaphysics and Unification 198
     References 201

Part I
Pluralism and Scientific Practice

Chapter 1
Beyond Placement Problems

Contemporary metaphysics is dominated by "placement problems".[1] Given that we live in a fundamentally physical world, how can we make sense of the existence of abstract objects, colors, indexicality, intentionality, macro-causation, mathematical entities, meaning, modality, normativity, ordinary objects, phenomenal consciousness, self-knowledge, social entities, and so on? Placement problems in contemporary metaphysics typically arise from three assumptions. First, a base of fundamental entities such as all physical objects, properties, and facts. Second, a supposedly non-fundamental entity that cannot be easily understood in terms of the base of fundamental entities. Third, the claim that we should be able to make sense of the latter in terms of the former. Metaphysicians usually acknowledge that placement problems can be solved in three incompatible ways:

(a) *Ontological Reduction*: We can solve placement problems by reducing non-fundamental entities to the base of fundamental entities. For example, one may assume that we can make sense of ordinary objects by arguing that they are nothing but highly complex arrangements of microphysical entities. This strategy only requires an *ontological* reduction but does not necessarily involve reductions in a more narrow *epistemic* sense. For example, non-reductive physicalism in philosophy of mind is based on the assumption that ontological reductions do not require reductive explanations.

(b) *Elimination*: We can also solve a placement problem by eliminating problematic entities from our ontologies. For example, some metaphysicians suggest that we do not have to reduce ordinary objects to microphysical objects because the former do not *really* exist. Another eliminative strategy solves the placement problem of moral truths by arguing that there are no moral truths. For example, error theorists assume that all moral statements are false while non-cognitivists suggest that moral statements should not be understood as propositions with

---

[1] The notion of placement problems has been popularized by Price (e.g. 2004, 2011). For a helpful discussion of the relation between placement problems and physicalism, see Stoljar (2010, 19–27).

© Springer International Publishing Switzerland 2015
D. Ludwig, *A Pluralist Theory of the Mind*, European Studies
in Philosophy of Science 2, DOI 10.1007/978-3-319-22738-2_1

truth values and therefore do not raise a placement problem of moral truths. One way or another, there is no placement problem for moral truths because there are no moral truths.

(c) *Extending the Base*: Finally, we can solve a placement problem by adding problematic entities to the base of fundamental entities. For example, this strategy is common among dualists in philosophy of mind who claim that a fundamental account of the world will include both fundamental physical and mental entities. Furthermore, metaphysical pluralists like Popper (1978) assume that the furniture of the universe also includes fundamental non-physical and non-mental entities such as abstract objects.

Although placement problems are ubiquitous in contemporary metaphysics, the mind-body problem is arguably the most visible of them. It is largely uncontroversial in contemporary metaphysics that we should keep the base of fundamental entities as small as possible (e.g. Chalmers 2012, 20; Kim 2008, 125). Ideally, we should solve all placement problems in a physicalist manner by only allowing physical entities in the base. This methodological specification renders (c) unattractive and leaves us with (a) and (b). Even philosophers who are generally optimistic about reductive or eliminative strategies, however, often acknowledge that reductive or eliminative accounts of phenomenal consciousness remain unsatisfying.[2] As a result, philosophers of mind have developed increasingly sophisticated variants of (a) – (c) without coming close to a consensus view regarding the ontological status of mental entities.

The aim of this book is to challenge common debates in philosophy of mind by challenging the entire framework of placement problems. I will argue that placement problems arise from the problematic assumption of exactly one fundamental ontology that specifies the base of fundamental entities. I will propose a pluralist alternative that takes the diversity of our conceptual resources and ontologies seriously by refusing to identify only one of them as truly fundamental. The starting point of my discussion is the plurality of ontologies in scientific practice. Not only can we describe the world in terms of physical, biological, or psychological ontologies, but philosophers of science routinely identify much more specific ontologies in each domain. For example, there is not one unified biological ontology but rather a diversity of fields with different ontological needs such as anatomy, conservation biology, ethnobiology, ethology, evolutionary developmental biology, genetics,

---

[2] Chalmers' *Constructing the World* (2012) provides helpful examples of this optimism as his scrutability framework is applicable to a large variety metaphysical positions. Chalmers' postulates a compact base of fundamental truths and argues that all non-fundamental truths must be scrutable from this base. Different bases will lead to different metaphysical positions. For example, Chalmers suggests a base that allows (P) physical and (Q) phenomenal truths but not aesthetic, intentional, or moral truths. Physicalists and idealists will postulate an even more restricted base that excludes either (P) or (Q). Every metaphysical position with a compact base of fundamental truths implies that the large majority of placement problems can solved through reduction or elimination. Given Chalmers' scrutability framework, the main claim of my book is that the inscrutability of phenomenal truths is not puzzling as neither science nor philosophy give us good reasons to believe in a compact base of fundamental truths.

genomics, metagenomics, molecular biology, neuropharmacology, neurophysiology, paleobiology, systems biology, and so on.[3]

Of course, there is an important difference between the *de facto* plurality of ontologies in scientific practice and any substantive philosophical pluralism. While the plurality of scientific ontologies is a largely uncontroversial observation, most philosophers will reject substantive forms of philosophical pluralism and insist on the need for ontological unification. The aim of this book is not to reject the value of ontological unification in science or even of variants of non-reductive integration that are ubiquitous in many areas of the life sciences (e.g. Mitchell 2003; Craver 2007; Brigandt 2010). Instead, I want argue that there is no reason to assume that ontological unification must be possible everywhere and that it is much more reasonable to consider the scope of ontological unification an open empirical question. While ontological unification is often valuable and important in scientific practice, unification efforts in science do not justify the metaphysical claim that everything has to be understood in terms of exactly one fundamental ontology that specifies the base of truly fundamental entities. Furthermore, I will argue that placement problems only occur if we presuppose this ideal of exactly one fundamental ontology and insist that everything else *has to* be understood in terms of it. Without this ideal, the scope of ontological unification turns out to be an open empirical question and there is no need to present unification failures as philosophically puzzling "placement problems". Finally, I will apply this general pluralist account to issues in philosophy of mind and argue that contemporary debates about the mind-body problem are built on this problematic framework of placement problems.

1.1 Varieties of Naturalism

Most philosophers of mind will be skeptical of my suggestion that we should approach the mind-body problem by rejecting the entire framework of placement problems. One way of articulating this skepticism is to point out that placement problems are motivated by a naturalism that is considered non-negotiable by many philosophers. A scientifically minded philosophy needs to "naturalize" entities such as consciousness, intentionality, or normativity. In order to naturalize these entities, however, we need to show how they fit in a fundamentally physical world.

---

[3]A note on terminology. Every account of ontological diversity in terms of different domains, fields, disciplines, levels, and so on is in risk of being misunderstood as suggesting neatly separated ontologies without any integration whatsoever. Mitchell (2003) correctly warns of an "isolationist pluralism" (cf. Van Bouwel 2014) and proposes an integrative alternative that allows for cross-disciplinary and cross-level integration of methods, models, and ontologies. Ruphy (2011) makes a similar point by proposing a "foliated pluralism" that stresses the transdisciplinary character of "styles of scientific thinking" and their ontologies. Any satisfying pluralism clearly has to leave room for various reductive and non-reductive forms of integration. Although it is often convenient to talk about ontologies being relative to "domains" or "fields", this is not meant to deny the complex entanglement of our conceptual resources and ontologies.

A rejection of placement problems is therefore tantamount to a rejection of naturalism and unacceptable for a philosophy that is informed by modern science.

While this objection incorporates the framework of placement problems in the very idea of naturalism, I assume that this is an unfortunate way of understanding naturalism. Of course, "naturalism" is a vague and ambiguous term and there is little point in a verbal dispute about its correct definition or true meaning. However, it will be helpful to distinguish between a common use of the term in philosophy of mind and the naturalistic methodology that I will adopt in my discussion of pluralism.[4] Naturalism in philosophy of mind often focuses on the alleged need to naturalize mental entities by understanding them in terms of more fundamental biological, functional, or even physical entities. In an important sense, this naturalism starts with a metaphysical picture (e.g. physicalism) and assumes that the main task of naturalistic philosophers is to find a place for unruly entities such as consciousness in this picture.

In contrast with this "naturalism of placement problems", I will adopt a naturalistic approach that starts with scientific practice instead of metaphysics. Instead of assuming that science needs to validate a presupposed metaphysical picture, this alternative form of naturalism suggests that metaphysics needs to adapt to the reality of scientific practice.[5] Given this "naturalism of scientific practice", the starting point of metaphysical inquiry will neither be a physicalist nor a dualist ontology but the breathtaking diversity of ontologies that we encounter in scientific practice. There can be little doubt that the recent history of science has not led to a unification but rather to a staggering explosion of ontologies and "kinds of things" (cf. Dennett 2013) that have become objects of scientific inquiry. Furthermore, there is little reason to believe that this trend is going to slow down (or even inverse) in the foreseeable future as the constant growth of ontologies is a direct consequence of the continuing specialization of science with vast amounts of new ontologies being created for highly specific purposes such as "an ontology for cell types" (Bard et al. 2005), "an ontology for biological function based on molecular interactions" (Karp 2000), "an ontology for detecting and resolving data and schema-level

---

[4] Of course, distinctions between different kinds of naturalism are common (cf. Horst 2009) and many philosophers have argued for accounts of naturalism that differ from a "naturalism of placement problems" in one way or another. See, for example, Bickle (2003), Brigandt (2011), Dupré (2010), Ladyman and Ross (2007), McDowell (2004), and Price (2011) for diverse and partly incompatible formulations of naturalisms that are independent of placement problems.

[5] In this sense, my project shares important assumptions with current accounts of "scientific metaphysics" (cf. Ross et al. 2013). At the same time, not all self-identified "scientific metaphysicians" will agree with my pluralist proposal and there remains room for substantive disagreement on the role of unification in science. For example, Ladyman and Ross (2007) present unification as "the touchstone of legitimate metaphysics" (Melnyk 2013, 82). My naturalist approach is arguably more radical in assuming that the role and importance of unification is also determined in scientific practice (cf. 11.2). While Ladyman and Ross correctly point out the importance of unification in some areas of physics, other research contexts involve more moderate forms of integration or even disunity and have little use for global unification ideals. It therefore seems odd to combine a "naturalism of scientific practice" with a demand for *global* unification instead of accepting that the role of unification varies in different research contexts.

semantic conflicts" (Ram and Park 2004), "an ontology for pharmaceutical ligands" (Schuffenhauer et al. 2002), "an ontology for the distribution of differences between RDF graphs" (Berners-Lee et al. 2004), and so on.

Even if this plurality of scientific ontologies is the starting point of a "naturalism of scientific practice", unification still often plays an important role in understanding the relation between ontologies. Furthermore, there are many forms of non-reductive ontological integration that specify relations and dynamics without aiming at ontological unification in a strict sense. A plurality of ontologies in scientific practice therefore does not imply a disunity of unrelated entities and research programs.[6]

The cognitive sciences provide a great example of the diverse relations between ontologies that range from traditional theory reductions and reductive explanations to clearly non-reductive models of interactions between entities on different levels of biological and cognitive organization (cf. Dale 2008; Dale et al. 2009). Furthermore, cognitive science is also full of substantive scientific explanations such as mechanistic explanations (e.g. Bechtel 2008; Craver 2007) that neither fit traditional reductionist nor antireductionist accounts. Mechanistic explanations share some features with traditional accounts of reduction (e.g. they appeal to the components of a mechanism in its explanation) but do not imply that knowledge about mechanisms can be derived from knowledge about its components.

There is an obvious tension between a naturalism of scientific practice that starts with the diversity of explanations and ontologies in cognitive science and common naturalistic projects in philosophy of mind. Placement problems arise from the ideal of global ontological unification and the assumption that we have to understand all non-fundamental entities in terms of a compact base of truly fundamental entities. However, a naturalism of scientific practice seems to provide little support for this idea of a compact base that is presupposed in the framework of placement problems. On the contrary, one may argue that the diversity of scientific explanations and ontologies requires us to adopt a pluralist picture that acknowledges the importance of unification in some areas of scientific practice but also points out the ubiquity of unproblematic and productive forms of disunity.

While pluralism and disunity have few friends in analytic metaphysics, they have become mainstream positions in philosophy of science. The pluralist mainstream in contemporary philosophy of science traces back at least to the 1970's and Suppes' landmark article on "The Plurality of Science" (1978). Suppes' main target is the

---

[6]Although I argue against the availability of global epistemic and ontological unification, science may turn out to be unified in some other and more moderate sense (cf. 2011). For example, Otto Neurath's (1935/1983 cf. Cartwright et al. 2008 and Symons et al. 2011) classical account of the unity of science does not include global reduction and is entirely compatible with my proposal of "naturalism of scientific practice". Given a Neurathian perspective, we should endorse the ideal of a scientific practice that is unified in terms of "encyclopedic integration" (cf. Pombo 2011; Andler 2011) even if epistemic and ontological reductions fail. Of course, it is hard to deny that scientific research can be described as unified in some senses and as disunified in other senses. General endorsements of the (dis)unity of science therefore always involve a rhetorical component and focus on certain aspects.

unity of science. First, he suggests that the ideal of unified science suffers from an outdated atomist picture of fundamental entities while in reality "we cannot have a reduction of a subject matter to the ultimate physical entities because we do not know what those entities are" (1978, 7). Second, Suppes argues that the development of computers has taught us that the same cognitive structures can be realized in physically radically different structures. Reflecting the state of computer science in the late 1970s, Suppes points out that "old computers using vacuum tubes and modern computers using semiconductors can execute exactly the same programs and can perform exactly the same tasks. The differences in physical properties are striking between these two generations of computers [and reduction] below the level of the concepts of information processing, seems wholly uninteresting and barren. Reduction to physical concepts is not only impractical but also theoretically empty" (1978, 7). Suppes concludes that we have to fundamentally rethink science as a diverse human activity that does not develop towards a unified picture of eternal truths: "No area of experience is totally and completely settled by providing a set of basic truths; but rather, we are continually confronted with new situations and new problems, and we bring to these problems and situations a potpourri of scientific methods, techniques, and concepts, which in many cases we have learned to use with great facility" (1978, 54).

Of course, pluralism in contemporary philosophy of science (cf. Ruphy 2013) is by no means a homogeneous movement[7] and debates have changed considerably since the 1970s. For example, the main targets of contemporary pluralism are not the unity of science or even a unified scientific language anymore but epistemological claims about reductive explanation and metaphysical claims about ontological unification. However, both Suppes and contemporary pluralists are crucially motivated by the diversity of scientific practice that does not seem to support the unification ideals of monist philosophers.

---

[7] The diversity of pluralisms in philosophy of science is nicely illustrated by the confusing diversity of pluralism labels such as "active normative epistemic pluralism" (Chang 2012), "anything goes pluralism" (Mitchell 2004) "conceptual pluralism" (Putnam 2004; Forrai 2001) "cognitive pluralism" (Horst 2007), "explanatory pluralism" (McCaunly and Bechtel 2001; Campaner 2014), "foliated pluralism" (Ruphy 2011), "integrative pluralism" (Mitchell 2003), "interactive pluralism" (Van Bouwel 2014), "isolationist pluralism" (Mitchell 2003), "metaphysical pluralism" (Cartwright 1999), "methodological pluralism" (Richardson 2009), "moderate/temporary pluralism" (Van Bouwel 2014), "naturalistic pluralism" (Polger 2007), "non-relatvisitc pluralism" (Van Bouwel and Weber 2008), "orientational pluralism" (Rescher 1978), "perspectival pluralism" (Giere 2006), "physicalist pluralism" (Shearmur 2009), "realist pluralism" (Chakravartty 2011), or "scientific pluralism" (Kellert et al. 2006). In the following, I will provide at least a rough taxonomy of pluralist positions by distinguishing between conceptual, epistemological, and metaphysical pluralism as well as horizontal and vertical pluralism.

1.2   Reduction and Reductive Explanation

Pluralism in philosophy of science seems to challenge placement problems and therefore also common formulations of the mind-body problem in a straightforward way. Philosophers of mind assume that the irreducibility of mental states constitutes a deep philosophical problem. In contrast, pluralism in philosophy of science seems to suggest that worries about irreducibility are the result of a misleading ideal of global unification that is not supported by our best understanding of science. If the scope of reductions is an open empirical question, then there is no reason to assume that everything has to be reduced to a fundamental physical level. Given these considerations, we can apply a preliminary "argument from scientific practice" to philosophy of mind:

1) Naturalism of scientific practice suggests that the scope of reductions is an open empirical question and that there is no reason to consider limits of reduction philosophically puzzling.
2) If there is no reason to consider limits of reduction philosophically puzzling, then there is also no reason to consider the irreducibility of mental entities puzzling.
3) Common formulations of the mind-body problem are based on the assumption that the irreducibility of mental entities such as phenomenal consciousness or intentionality constitutes a deep philosophical problem.
∴ Naturalism of scientific practice undermines common formulations of the mind-body problem.

Somewhat surprisingly, this line of argument has largely failed to make an impact on debates in philosophy of mind. While pluralism, disunity, and "philosophy of scientific practice" have become mainstream positions in post-Kuhnian philosophy of science, philosophers of mind tend to be unimpressed by these developments and stick with the assumption that explanatory gaps in philosophy of mind are something unique and deeply puzzling.

An important exception is Steven Horst's *Beyond Reduction* (2007), which argues that "philosophy of mind at the turn of the millennium is, as it were, one of the last bastions of 1950's philosophy of science" (2007, 4), and that we can solve issues in philosophy of mind by taking pluralist developments in philosophy of science more seriously. As Horst puts it: "But of course, if this is so – if explanatory gaps are commonplace – then it is not clear why the psychological gaps ought to be such a big deal. There is obviously an important question of why there are, in general, such gaps in our understanding. But thus far we have no reason to suppose that the reasons for the mind-brain gap are fundamentally different from those for, say, the evolution– molecular genetics" (2007, 84).

Although Horst argues for radical implications of scientific pluralism for the mind-body problem, few philosophers of mind seem to be inclined to follow. One way of interpreting this situation is to argue that philosophers of mind are simply unaware of the radical implications of contemporary philosophy of science.

Horst provides an example of this strategy by arguing that "the entire problematic [of psychophysical explanatory gaps] is an artifact of an erroneous view in the philosophy of science. The crucial error is to assume that intertheoretic reductions *are* the norm" (2007, 4).

While I agree with Horst that there is a disconnect between large parts of philosophy of science and philosophy of mind (and analytic metaphysics in general), philosophers of mind can object that explanatory gap problems reappear even under the assumption of a reasonably strong notion of disunity of science. Indeed, philosophers of mind have not been completely oblivious to developments in philosophy of science since the 1950s and the non-availability of traditional theory reductions is not really news to them given the well-known debates about multiple realization and the autonomy of the special sciences (Fodor 1974; cf. Otero 2013; van Riehl 2014). However, disunity in the sense of Fodor does not imply a pluralist theory that rejects placement problems altogether but rather leads to a more modest idea of reductive explanation. As Fodor puts it: "It seems to me (to put the point quite generally) that the classical construal of the unity of science has really misconstrued the goal of scientific reduction. The point of reduction is not primarily to find some natural kind predicate of physics co-extensive with each natural kind predicate of a reduced science. It is, rather, to explicate the physical mechanisms whereby events conform to the laws of the special sciences" (Fodor 1974, 107).

Fodor's comments provide a helpful starting point for discussions of scientific explanations that are weaker than traditional theory reductions but preserve the idea of an explanatory disanalogy between innocent entities of the so-called special sciences and troubling entities of placement problems. Even if the special sciences are autonomous in the sense of Fodor, one may still insist that we can explain most entities of the special sciences in terms of fundamental physical entities. However, even more moderate accounts of reductive explanation fail in the case of placement problems such as phenomenal consciousness and therefore lead to genuine philosophical problems. Arguably, popular formulations of explanatory gap arguments (e.g. Levine 1983) are based on exactly this disanalogy as they claim that not only traditional theory reductions but even more moderate accounts of reductive explanation fail in the case of consciousness. Furthermore, there is an abundance of literature on more moderate proposals such as Kim's (2005, 2008) "functional reduction", Chalmers and Jackson's (2001) "reductive explanations" and Chalmers' (2012) more recent "scrutability" that are compatible with a rejection of unification ideals that are based on traditional theory reductions.

Given this distinction between reduction and more moderate accounts of reductive explanation (or functional reduction, scrutability…), it seems that proponents of explanatory gap problems cannot only respond to Horst's challenge but also to my preliminary formulation of an argument from scientific practice. If reduction is understood in a narrow sense as traditional theory reduction, they will reject the third premise of the argument by pointing out that current accounts of explanatory gaps are not based on the failure of theory reductions but on the failure of reductive explanations. If reduction is understood in a broader sense that includes more moderate accounts of reductive explanations, they will reject the first premise by arguing that contemporary cognitive science actually suggests that reductive explanations are often successful even if traditional theory reductions fail.

This appeal to "reductive explanations without reduction" (Chalmers 2012, 309) can be further motivated by case studies from cognitive science. For example, the psychology of learning is not reducible to a neuroscientific level in the sense of classical theory reductions. The most obvious reason is that learning psychology does not come with laws akin to those in classical cases of theory reduction. Furthermore, learning processes such as habituation, sensitization, and classical conditioning are common among a large variety of species and are arguably not always realized by the same neural mechanisms.[8]

Even if classical theory reductions fail in the case of learning psychology, it still seems plausible that scientists can explain learning processes in terms of neural processes in a more moderate sense. For example, habituation is a process in which a system comes to ignore an unimportant stimulus. In many organisms with comparably simple nervous systems such as *Aplysia*, neuroscience provides very detailed accounts of the molecular mechanisms that underlie habituation. Neuroscientists can explain on a molecular level why certain stimuli become ignored over time by describing the decreasing amount of the neurotransmitter glutamate at the interface between sensory neurons and motor neurons. It therefore seems at least plausible that neuroscience can explain habituation in *Aplysia* even if traditional accounts intertheoretic reduction fail.

It is not difficult to see how the example of habituation can be contrasted with phenomenal states such as pain. In the case of pain, neuroscientists can also provide a detailed account of neural and molecular mechanisms but they will not offer an explanation of pain in the same sense as they offer an explanation of habituation in *Aplysia*. In the case of the former, the question *why* an organism feels pain remains open while there is no comparable question in the case of habituation in *Aplysia*. Therefore, there remains a unique explanatory gap between phenomenal states and neural states that creates a placement problem: We do not know how phenomenal states fit in the physical world.

The moral of this example seems to be that a pluralist challenge of placement problems cannot be limited to discussions of traditional theory reductions. Instead, a successful challenge needs to defend a more comprehensive pluralism that does not only challenge the expectation of global availability of theory reductions but also of reductive explanations. As long as pluralism only challenges theory reductions, philosophers of mind can insist on a fundamental disanalogy between the large number of innocent scientific entities that are explicable in terms of a fundamental physical ontology and the small number of puzzling entities that constitute placement problems.

---

[8] For a more careful discussion of this example and references, see 6.6. The multiple realization of elementary learning is not as uncontroversial as one may expect (cf. Bickle 2003, Sect. 5.2, Aizawa 2007). However, not much depends on this in the present context.

1.3   Epistemological, Metaphysical, and Conceptual Pluralism

A rejection of traditional theory reductions is not sufficient for a rejection of explanatory gap problems in philosophy of mind as one can appeal to more moderate notions of reductive explanation. At the same time, pluralists can respond by endorsing a pluralism that extends from theory reduction to reductive explanation. Even if we accept that examples such as learning in *Aplysia* show that reductive explanations without theory reductions are possible, we do not have to assume that reductive explanations will be successful everywhere. Reductive explanations are clearly important in some research contexts, but that does not mean that we have to endorse a global reductivism. In strict analogy to the challenge of theory reductions, a naturalism of scientific practice seems to suggest that we should treat the scope of both reductions *and* reductive explanations an open empirical question. We do not know whether future neuroscience will eventually provide an explanation of phenomenal consciousness that satisfies standard models of reductive explanations in philosophy of mind. However, there is no reason to treat this open empirical question as a deep philosophical problem as long as we do not presuppose a reductivism that claims that reductive explanations must be successful everywhere. It therefore also seems that we can reformulate the argument from scientific practice by substituting "reduction" with "reductive explanation":

1) Naturalism of scientific practice suggests that the scope of reductive explanations is an open empirical question and that there is no reason to consider limits of reductive explanations philosophically puzzling.
2) If there is no reason to consider limits of reductive explanation philosophically puzzling, then there is also no reason to consider explanatory gaps in philosophy of mind philosophically puzzling.
3) Common formulations of the mind-body problem are based on the assumption that explanatory gaps in philosophy of mind constitute a deep philosophical problem.
∴ Naturalism of scientific practice undermines common formulations of the mind-body problem.

So far, my discussion has been built on the claim that scientific practice gives us no reason to believe that reductions or reductive explanations must be successful everywhere. However, there may still be other – e.g. metaphysical – reasons to insist on reductivism. For example, one may object that a pluralist appeal to the diversity of scientific practice does not lead to a coherent and substantive philosophical pluralism. More specifically, one may argue that any attempt to formulate a substantive pluralism will lead to a dilemma between a weak epistemological pluralism that does not solve placement problems and an overly strong and highly implausible metaphysical pluralism.

On the one hand, one may present scientific pluralism as a merely epistemological pluralism that is concerned with the plurality of scientific explanations but still endorses the ideal of global ontological unification. In this case, one can object that scientific pluralism does not actually lead to a novel position in philosophy of mind but to common variants of non-reductive physicalism. Furthermore, this merely epistemological pluralism will face the same problems as non-reductive physicalism in debates about the mind-body problem. Most importantly, it seems that the assumption of exactly one fundamental physical ontology renders the unavailability of reductive explanations mysterious. How is it possible that all entities are fundamentally physical entities if some of them are not explicable in terms of physics?[9]

On the other hand, one can also propose a more radical metaphysical pluralism that rejects physicalism by insisting on a diversity of non-physical entities. However, one may object that this is a highly implausible metaphysical picture that faces the same problems as traditional variants dualism such as problems of the causal efficiency of non-physical entities. For example, consider Popper's metaphysical pluralism of "three worlds" (1978) that includes a traditional interactionist mind-body dualism and the assumption of additional non-physical and non-mental entities. Popper's metaphysical pluralism does not only run into the same problems as an interactionist dualism but even creates further problems with the assumption of non-physical and non-mental entities such as abstract objects. The example therefore suggests that a strong metaphysical pluralism will not be an improvement over well-known variants of dualism and will have to be combined with a highly problematic interactionist or epiphenomenalist theory of the relation between physical and non-physical entities.

To sum up, one may object that every pluralist account leads to a dilemma of a weak epistemological pluralism and an overly strong metaphysical pluralism. An epistemological pluralism will face the same problems as well-known variants of non-reductive physicalism and a metaphysical pluralism will face the same problems as well-known variants of dualism. One way or another, pluralism does not add anything novel or attractive to debates about the mind-body problem.

Although this dilemma constitutes an important challenge for a pluralist theory of mind, I will argue that it can be met and I will propose a conceptual pluralism that avoids both horns of the dilemma. Engagement with the diversity of ontologies in scientific practice suggests a pluralism that is considerably stronger than a merely epistemological pluralism but does not imply a traditional dualist picture of metaphysically distinct realms of reality. Instead, I will argue that conceptual pluralism implies a plurality of ontologies that are shaped by our diverse conceptual resources. This conceptual pluralism differs from a merely epistemological pluralism by rejecting the idea of one fundamental ontology and differs from an overly strong

---

[9] This does not mean that an epistemological pluralism is of no value whatsoever. For example, many pluralist accounts in philosophy of neuroscience are only concerned with epistemological and methodological issues and avoid ontological debates (e.g. Craver 2007, 13). This strategy is well-justified in the context of a philosophy of neuroscience that is interested in the structure of neuroscientific explanations but not concerned with metaphysical issues in philosophy of mind. Problems arguably only occur, if a "merely epistemological pluralism" is presented as a solution of traditional problems in philosophy of mind.

metaphysical pluralism by interpreting ontological diversity in terms of diverse conceptual resources instead of metaphysically distinct realms of reality.

In order to provide a preliminary illustration of this idea of conceptual pluralism, consider a few examples of ontological controversies in the empirical sciences (cf. Ludwig 2013, 2014). In biology, different species concepts provide a convenient case study of how different conceptual choices lead to different biological ontologies. For example, the so-called biological, ecological, and morphological species concepts imply the existence of different biological entities and in this sense different biological ontologies. A strong pluralist interpretation of the species debate therefore not only implies an epistemological pluralism but also an ontological pluralism that rejects the idea of exactly one fundamental biological ontology. However, this ontological pluralism is not simply an extension of dualism such as Popper's theory of "three worlds". Biological, ecological, and morphological species concepts do not refer to metaphysically distinct realms of reality but describe the same biological reality in terms of different ontologies.

I will argue that analogous cases of ontological pluralism are ubiquitous in science. For example, contemporary psychiatry comes with a variety of ontologies that postulate different mental disorders. Again, it would be odd to claim that different psychiatric ontologies describe distinct realms of reality in the sense of an ambitious metaphysical pluralism. Instead, a conceptual pluralist will suggest that we can describe the human mind in variety of ways that imply a variety of psychiatric ontologies. In order to further illustrate this distinction between conceptual pluralism and an ambitious metaphysical pluralism, consider issues of causation. An ambitious metaphysical pluralism will face the same kind of concerns regarding overdetermination and mental causation as a dualist (e.g. Kim 2005, cf. Chapter 10). My examples of ontological plurality in scientific practice do not raise analogous worries. For example, a behavior that is caused by "hysteria" according to the DSM-II can be caused by "somatoform disorder" according to the DSM-III. However, no one would assume that there is a mysterious causal overdetermination through hysteria and somatoform disorder. Instead, it seems obvious that psychiatrists describe mental phenomena in terms of different psychiatric ontologies but do not refer to metaphysically distinct entities that could compete for causal relevance. I argue that this argument extends to conceptual pluralism in general. Limits of ontological unification do not imply metaphysically distinct realms of reality and therefore do not imply a mind-body dualism or a problem of mental causation.

1.4 The Argument in a Nutshell

Conceptual pluralism avoids the dilemma of a merely epistemological pluralism and a strong metaphysical pluralism by arguing for a plurality of ontologies that are shaped by different conceptual resources. Contrary to a merely epistemological pluralism, conceptual pluralism rejects the ideal of global ontological unification and the ideal of exactly one fundamental ontology. Contrary to a strong metaphysical pluralism, conceptual pluralism interprets this ontological diversity in terms of diverse conceptual resources instead of metaphysically distinct realms of reality.

The goal of this book is to develop a "pluralist theory of the mind" on the basis of this conceptual pluralism. I will argue that conceptual pluralism challenges both the general metaphysical obsession with placement problems as well as more specific debates about explanatory gaps in philosophy of mind. Scientists describe reality in terms of different ontologies and we should treat issues of ontological unification as open empirical questions. While ontological unification can be an important goal in scientific practice, there is no reason to assume that ontological unification will (or even must) be successful *everywhere* in the sense that everything can be explained in terms of one fundamental physical ontology.

It is helpful to distinguish between three premises of my argument for a pluralist theory of mind. The first premise is a general conceptual pluralism that insists on a plurality of equally fundamental ontologies. Even if the *de facto* plurality of ontologies in scientific practice is uncontroversial, a pluralism of *equally fundamental* ontologies is highly controversial and requires justification. My discussion of conceptual pluralism will proceed in two steps. First, I discuss the idea of equally fundamental ontologies in philosophy on the basis of Putnam's theory of conceptual relativity and current metaontological debates. In a second step, I argue that the coexistence of different but equally fundamental ontologies is ubiquitous in science and should be considered largely uncontroversial in the framework of a naturalism of scientific practice.

The second premise of my argument is based on the assumption that issues of ontological priority are closely related to debates about reductive explanation. The assumption of exactly one fundamental (e.g. physical) ontology justifies the reductivist claim that everything must be – at least in principle – explicable in terms of a fundamental (e.g. physical) ontology. An ontological pluralism undermines this motivation of reductivism and instead suggests that the scope of reductive explanations is an open empirical question. Although we should not reject the possibility of reductive explanations of phenomenal consciousness and other entities, there is also no reason to assume that reductive explanation must be successful everywhere. Failed reductions and reductive explanations are philosophically not more troubling than successful reductions and reductive explanations.

According to the third premise of my argument, common formulations of the mind-body problem presuppose a reductivism that is motivated by the idea of one fundamental ontology. For example, consider explanatory gap arguments in philosophy of mind. A conceptual pluralist may accept an explanatory gap in the sense that we cannot reductively explain phenomenal states in terms of biological or physical states, but she will argue that this explanatory gap is only puzzling under the assumption of a dubious global reductivism.

My goal is to offer a comprehensive presentation and defense of this argument, with each of the following parts devoted to one of its premises. Part II will be concerned with the general assumption of a plurality of equally fundamental ontologies, Part III will spell out the relationship between ontological pluralism and nonreductivism, and Part IV will apply this framework to the mind-body problem.

References

Aizawa, Kenneth. 2007. The Biochemistry of Memory Consolidation: A Model System for the Philosophy of Mind. *Synthese* 155 (1): 65–98.

Andler, Daniel. 2011. Unity Without Myths. In *Otto Neurath and the Unity of Science*, eds. Symons, John, Olga Pombo, and Juan Manuel Torres 129–144. Berlin: Springer.

Bard, Jonathan, Seung Y. Rhee, and Michael Ashburner. 2005. An Ontology for Cell Types. *Genome Biology* 6 (2): R21.

Bechtel, William. 2008. *Mental Mechanisms: Philosophical Perspectives on Cognitive Neuroscience*. New York: Taylor & Francis.

Berners-Lee, Tim, and Dan Connolly. 2004. Delta: An Ontology for the Distribution of Differences Between RDF Graphs. *MIT Computer Science and Artificial Intelligence Laboratory.*

Bickle, John. 2003. *Philosophy and Neuroscience: A Ruthlessly Reductive Account.* Boston: Kluwer.

Van Bouwel, Jeroen. 2014. Pluralists about pluralism? Different versions of explanatory pluralism in psychiatry. *New Directions in the Philosophy of Science,* eds. Galavotti et al., 105–119. Berlin: Springer.

Van Bouwel, Jeroen, and Erik Weber. 2008. A Pragmatist Defense of Non-Relativistic Explanatory Pluralism in History and Social Science. *History and Theory* 47(2): 168–182.

Brigandt, Ingo. 2010. Beyond Reduction and Pluralism: Toward an Epistemology of Explanatory Integration in Biology. *Erkenntnis* 73 (3): 295–311.

Brigandt, Ingo. 2011. Natural Kinds and Concepts: A Pragmatist and Methodologically Naturalistic Account. In *Pragmatism, Science and Naturalism,* eds. Jonathan Knowles and Henrik Rydenfelt, 171–196. Frankfurt am Main: Peter Lang.

Campaner, Raffaella. 2014. Explanatory Pluralism in Psychiatry: What Are We Pluralists About, and Why? In *New Directions in the Philosophy of Science,* eds. Galavotti et al., 87–103. Berlin: Springer.

Cartwright, Nancy. 1999. *The Dappled World: A Study of the Boundaries of Science*. Cambridge: Cambridge University Press.

Cartwright, Nancy, Lola Fleck, and Thomas E. Uebel. 2008. *Otto Neurath: Philosophy between science and politics*. Cambridge: Cambridge University Press.

Chalmers, David. 2012. *Constructing the World*. Oxford: Oxford University Press.

Chalmers, David J., and Frank Jackson. 2001. Conceptual Analysis and Reductive Explanation. *Philosophical Review* 110: 315–60.

Chakravartty, Anjan. 2011. Scientific Realism and Ontological Relativity. *Monist* 94(2): 157–180.

Chang, Hasok. 2012. *Is Water H2O?: Evidence, Realism and Pluralism*. Berlin: Springer.

Craver, Carl F. 2007. *Explaining the Brain*. Oxford: Oxford University Press.

Dale, Rick. 2008. The Possibility of a Pluralist Cognitive Science. *Journal of Experimental and Theoretical Artificial Intelligence* 20(3): 155–179.

Dale, Rick., Dietrich, Eric., and Chemero, Anthoney 2009. Explanatory Pluralism in Cognitive Science. *Cognitive Science* 33 (5): 739–742.

Dennett, Daniel. 2013. Kinds of Things. Bestiary of the Manifest Image. In *Scientific Metaphysics*, eds. Don Ross, James Ladyman, and Harold Kincaid, 96–107. Oxford: Oxford University Press.

Dupré, John. 2010. How to Be Naturalistic Without Being Simplistic in the Study of Human Nature. In *Naturalism and Normativity*, eds. Mario De Caro and David Macarthur, 289–303. New York: Columbia University Press.

Fodor, Jerry A. 1974. Special Sciences (or: The Disunity of Science as a Working Hypothesis). *Synthese* 28 (2): 97–115.

Forrai, Gábor. 2001. *Reference, Truth and Conceptual Schemes: A Defense of Internal Realism*. Berlin: Springer.

Giere, Ronald N. 2006. *Scientific Perspectivism*. Chicago: University of Chicago Press.

Horst, Steven W. 2007. *Beyond Reduction: Philosophy of Mind and Post-reductionist Philosophy of Science*. Oxford: Oxford University Press.

Giere, Ronald N. 2009. Naturalisms in Philosophy of Mind. *Philosophy Compass* 4 (1): 219–254.

Karp, Peter D. 2000. An Ontology for Biological Function Based on Molecular Interactions. *Bioinformatics* 16 (3): 269–85.

Kellert, Stephen H., Helen E. Longino, and C. Kenneth Waters. 2006. *Scientific Pluralism*. Vol. 19. Minneapolis: University of Minnesota Press.

Kim, Jaegwon. 2005. *Physicalism, or Something Near Enough*. Princeton: Princeton University Press.

Kim, Jaegwon. 2008. Reduction and Reductive Explanation: Is One Possible Without the Other. In *Being Reduced*, eds. Jakob Hohwy and Jesper Kallestrup, 93–114. Oxford: Oxford University Press.

Ladyman, James, and Don Ross. 2007. *Every Thing Must Go: Metaphysics Naturalized*. Oxford: Oxford University Press.

Levine, Joseph. 1983. Materialism and Qualia: The Explanatory Gap. *Pacific Philosophical Quarterly* 64 (4): 354–61.

Ludwig, David. 2013. Hysteria, Race, and Phlogiston. A Model of Ontological Elimination in the Human Sciences. *Studies in History and Philosophy of Science Part C: Studies in History and Philosophy of Biological and Biomedical Sciences* 45: 67–77.

Ludwig, David. 2014. Disagreement in Scientific Ontologies. *Journal for General Philosophy of Science* 45 (1): 1–13.

McCauley, Robert N., and William Bechtel. 2001. Explanatory Pluralism and Heuristic Identity Theory. *Theory and Psychology* 11 (6): 736–760.

McDowell, John. 2004. Naturalism in the Philosophy of Mind. In *Naturalism in Question*, eds. Mario De Caro and David Macarthur, 91–106. Harvard: Harvard University Press.

Melnyk, Andrew. 2013. Can Metaphysics Be Naturalized? And If so, How? In *Scientific Metaphysics*, eds. Don Ross, James Ladyman, and Harold Kincaid, 79–95. Oxford: Oxford University Press.

Mitchell, Sandra D. 2003. *Biological Complexity and Integrative Pluralism*. Cambridge: Cambridge University Press.

Mitchell, Sandra D. 2004. Why integrative pluralism? *E:CO* 6 (1/2): 81–91.

Neurath, Otto. 1935/1983. The unity of science as a task. In *Philosophical Papers 1913–1946*, 115–120. Berlin: Springer.

Otero, Juan Diego Morales. 2013. Fodor y Kim en torno a la posibilidad de las ciencias especiales, la realizabilidad múltiple y el reduccionismo. *Revista Colombiana de Filosofía de la Ciencia* 13 (27): 63–84.

Polger, Thomas W. 2007. Some Metaphysical Anxieties of Reductionism. *The Matter of the Mind: Philosophical Essays on Psychology, Neuroscience and Reduction*, eds. Maurice Kenneth Davy Schouten, and Huibert Looren de Jong, 51–75. Cambridge, Mass.: MIT Press.

Pombo, Olga. 2011. Neurath and the Encyclopaedic Project of Unity of Science In *Otto Neurath and the Unity of Science*, eds. Symons, John, Olga Pombo, and Juan Manuel Torres, 59–70. Berlin: Springer.

Popper, Karl Raimund. 1978. *Three Worlds. The Tanner Lecture on Human Values*. Minneapolis: University of Michigan.

Price, Huw. 2004. Naturalism Without Representationalism. In *Naturalism in Question*, eds. Mario De Caro and David Macarthur, 71–88. Harvard: Harvard University Press.

Price, Huw. 2011. *Naturalism Without Mirrors*. Oxford: Oxford University Press.

Putnam, Hilary. 2004. *Ethics Without Ontology*. Harvard: Harvard University Press.

Ram, Sudha, and Jinsoo Park. 2004. Semantic Conflict Resolution Ontology (SCROL): An Ontology for Detecting and Resolving Data and Schema-level Semantic Conflicts. *Knowledge and Data Engineering, IEEE Transactions On* 16 (2): 189–202.

Rescher, Nicholas. 1978. Philosophical Disagreement: An Essay Towards Orientational Pluralism in Metaphilosophy. *The Review of Metaphysics* 32 (2): 217–251.

Richardson, Robert C.. 2009. Multiple Realization and Methodological Pluralism. *Synthese* 167 (3): 473–92.

van Riel, Raphael. 2014. *The Concept of Reduction*. Berlin: Springer 2014.

Ross, Don, James Ladyman, and Harold Kincaid. 2013. *Scientific Metaphysics*. Oxford: Oxford University Press.

Ruphy, Stéphanie. 2011. From Hacking's Plurality of Styles of Scientific Reasoning to 'foliated' Pluralism: A Philosophically Robust Form of Ontologico-Methodological Pluralism. *Philosophy of Science* 78 (5): 1212–22.

Ruphy, Stéphanie. 2013. *Pluralismes Scientifiques. Enjeux Épistémiques et Métaphysiques*. Paris: Hermann.

Schuffenhauer, Ansgar, Jürg Zimmermann, Ruedi Stoop, Jan-Jan van der Vyver, Steffano Lecchini, and Edgar Jacoby. 2002. An Ontology for Pharmaceutical Ligands and Its Application for in Silico Screening and Library Design. *Journal of Chemical Information and Computer Sciences* 42 (4): 947–55.

Shearmur, Jeremy. 2009. Our Place in Nature. McHenry, L. (Ed.). (2009). *Science and the pursuit of wisdom: Studies in the philosophy of Nicholas Maxwell*. Frankfurt: Ontos Verlag.

Stoljar, Daniel. 2010. *Physicalism*. New York: Routledge.

Suppes, Patrick. 1978. The Plurality of Science. *PSA: Proceedings of the Biennial Meeting of the Philosophy of Science Association*, 3–16.

Symons, John, Olga Pombo, and Juan Manuel Torres. 2011. *Otto Neurath and the unity of science*. Berlin: Springer.

Chapter 2
A Historical Diagnosis

Technicalities aside, the basic idea of my pluralist proposal is quite simple. We can describe reality in terms of vastly different conceptual resources. For example, we can describe ourselves as physical objects, biological organisms, intelligent beings, moral agents, and so on. Furthermore, neither of these conceptual resources has a homogenous structure. For example, biological accounts include conceptual resources from anatomy, conservation biology, ecology, ethology, evolutionary biology, molecular biology, morphology, physiology, population genetics, and so on. All of this is entirely unproblematic until metaphysicians come along and insist that we have to understand our complex conceptual resources on the basis of a compact base of truly fundamental entities. In contrast with this assumption, I want to suggest that we should not presuppose a compact base of truly fundamental entities. Although reductions and reductive explanations often play an important role in scientific practice, there is little reason to believe that we end up with a compact base that only includes one or two types of entities as suggested by the common metaphysical options of physicalism, dualism, and idealism.

While this pluralist proposal shares important assumptions with mainstream positions in philosophy of science, it constitutes a fringe position in philosophy of mind where the metaphysical ideal of a compact base of fundamental entities remains largely unquestioned. Most philosophers of mind consider physicalism and dualism as their only credible options: the base of fundamental entities either includes only physical entities or it includes both physical and mental entities. While the ideal of a compact base of fundamental entities has become a largely unquestioned starting point in philosophy of mind, I want to argue that it not only stands in the way of a convincing metaphysics of mind but also comes at the price of ignorance towards large parts of the history of philosophy of mind.

Philosophy of mind was not always concerned with placing the mind in a fundamentally physical world and I want to provide at least the outlines of a historical diagnosis. More specifically, I want to suggest that the assumption of a compact base of fundamental entities constitutes a core difference between modern analytic

© Springer International Publishing Switzerland 2015
D. Ludwig, *A Pluralist Theory of the Mind*, European Studies
in Philosophy of Science 2, DOI 10.1007/978-3-319-22738-2_2

philosophy of mind and many philosophical theories of the mind until the second half of the twentieth century.

Although analytic philosophy of mind has developed a highly productive research program, it has also narrowed down the scope of acceptable metaphysical positions. In a somewhat loose analogy to Kuhn (1962), analytic philosophy of mind has adopted many characteristics of a "normal science" that is highly productive on a background of shared core beliefs.[1] Normal science in the Kuhnian sense is an important aspect of scientific practice as it allows stable debates that progress in solving of increasingly specialized problems that he also describes as "puzzle solving" (1962, 35). Philosophy of mind since the second half of the 20th is arguably as close to "normal science" as philosophy can get. First, post-war philosophy of mind has developed a technical apparatus that matches many empirical sciences in terms of specialization and as illustrated by extensive debates about conceivability, embodiment, emergence, externalism, identity, language of thought, mental causation, modularity, multiple realization, necessity, reduction, reductive explanation, supervenience, and so on. Furthermore, there can be little doubt that this increasing specialization has led to tremendous progress in terms of Kuhnian puzzle solving. For example, consider the development from Davidson's vague statement "that mental characteristics are in some sense dependent, or supervenient, on physical characteristics" (1970, 214) to contemporary debates about countless varieties of (e.g. weak, strong, metaphysical, logical, individual, regional, global, mereological...) supervenience that have greatly clarified the implications of supervenience claims and are probably as close as one can get to the idea of scientific knowledge accumulation in philosophy.

Progress in a Kuhnian normal science comes at the price of an exclusion of alternative positions that do not share basic background assumptions. Again, the analogy between philosophy of mind and an increasingly specialized normal science can be helpful as philosophers of mind ignore a large variety of contemporary and historical positions from their debates. In some cases such as idealism and constructivism, most philosophers of mind will be happy to admit an exclusion on the basis of largely unquestioned background assumptions. For example, Goodman's (1978) *Ways of Worldmaking* may provide an obvious solution to explanatory gap problems in philosophy of mind (no need worry about the irreducibility of consciousness if physics only describes one of our many man-made worlds) but remains excluded from philosophy of mind due to the background assumptions that any convincing theory of the mind needs to be compatible with a minimal realism. At the same time, I will argue that contemporary philosophy of mind also excludes many alternatives that do not violate largely uncontroversial background assumptions such as a minimal realism. In the following, I will illustrate this claim by focusing on a historically influential but largely forgotten position that shares crucial assumptions with my proposal of conceptual pluralism: psychophysical parallelism in the tradition of Moritz Schlick, the founding father of the Vienna Circle.

---

[1] Of course, the characterization of philosophy of mind as "normal science" should not be taken too seriously and follows a general trend to use Kuhn's terminology to describe general features of academic discourse dynamics. Furthermore, it leaves the question open whether philosophers should aim at research that resembles "normal science" in the empirical sciences.

2.1 Schlick's Challenge

Contemporary textbooks in philosophy of mind introduce psychophysical parallelism as a bizarre metaphysical theory.[2] Most commonly associated with Leibniz' pre-established harmony, psychophysical parallelism is understood as the claim that the mental and material realms are not only ontologically but also causally separated. God created the mental and the material in a perfect harmony, which neither requires nor allows causal interaction. Just as two ideal clocks can be perfectly synchronized without causally influencing one another, the mind and the body run alongside each other without any interaction.

Given this characterization, Moritz Schlick seems to be an unlikely proponent of psychophysical parallelism. However, in a letter to Ernst Cassirer, he endorsed psychophysical parallelism as an entirely satisfactory approach to the mind–body problem: "The psychophysical parallelism in which I firmly believe [...] is a harmless parallelism of two differently generated concepts. Many oral discussions on this point have convinced me (and others) that this way we can get rid of the psychophysical problem once and for all" (1927). Schlick's endorsement of psychophysical parallelism is question-begging. How can he consider psychophysical parallelism "harmless," and how can he argue that it is only a parallelism of "two differently generated concepts?" The obvious answer is that Schlick's theory has little to do with the common textbook version of psychophysical parallelism. In fact, Schlick's psychophysical parallelism has little to do with *any* of the standard positions one finds in a philosophy of mind textbook.

Schlick's most detailed discussion of the mind–body problem comes in his *General Theory of Knowledge*, which acknowledges it as a "problem that, since the times of Descartes, has been at the center of all metaphysics," but still argues that it owes its "existence to a mistaken formulation of the issue" (1918/1974, 289). Given Schlick's role in the neopositivist movement, it might not come as a surprise that he rejected the mind-body problem as ill-conceived. Schlick's psychophysical parallelism, however, is not a consequence of verificationism or anti-metaphysical sentiment. On the contrary, Schlick's *General Theory of Knowledge*, which was first published in 1918, endorses a resolute realism that would appear as hopelessly metaphysical by the standards of later neopositivistic doctrines.

According to Schlick, both the mental and the physical are unproblematic aspects of reality. However, contrary to dualism and the common textbook interpretations of parallelism, the mental and physical are not understood as two metaphysically distinct aspects of reality. Instead, Schlick insists that his psychophysical parallelism "is an epistemological parallelism between a psychological conceptual system on the one hand and a physical conceptual system on the other. The 'physical world' *is* just the world that is designated by means of the system of quantitative concepts

---

[2] Examples include Braddon-Mitchell and Jackson (2007, 17–18), Campbell (2005, 28–30), and Heil (2012, 31–33). Braddon-Mitchell and Jackson nicely summarize the current attitude: "There are, as far as we know, no parallelists left. So we'll pass over this implausible view" (18).

of the natural sciences" (1918/1974, 301). Whereas dualists assume an ontological gap, Schlick argues for a conceptual gap. "There is only *one* reality," (1918/1974, 244) but this reality can be described in terms of different conceptual systems.

Schlick's conceptual parallelism would provide no challenge to contemporary philosophy of mind, if it were only directed against dualism. On the contrary, much of what he says seems to resemble contemporary non-reductive physicalisms that insist on differences between physical and phenomenal concepts (e.g. Loar 1990; Hill and McLaughlin 1999). However, Schlick goes further than that and presents materialism and other variants of "metaphysical monism" as equally flawed. According to Schlick, we have to take the plurality of conceptual systems seriously. A materialist who assumes that only physics describes what fundamentally exists makes the same mistake as a "psychomonist" who considers only the mental to be real or fundamental. "Earlier we were obligated most emphatically to reject the mistaken idea that a different kind or a different degree of reality must be ascribed to these two groups of reality [the mental and the extra-mental], that one group is to be characterized as merely an 'appearance' of the other. On the contrary, they are all to be regarded as, so to speak, of equal value" (1918/1974, 292). The assumption that *only* physical concepts provide a truly fundamental description of reality is the fatal flaw of materialism, "which in its admiration for the solid reality of physical objects simply forgets that there also exists a real world of consciousness or believes that it may be treated as a quantité négligeable" (1918/1974, 234).

Schlick's psychophysical parallelism presents a unique challenge to common formulations of the mind-body problem that only consider materialism and dualism. Contrary to the large majority of contemporary philosophers of mind, Schlick assumes that *both* materialism and dualism are flawed. He challenges both metaphysical doctrines by suggesting that we can describe the world in terms of different but equally fundamental conceptual systems. Materialism is wrong in assuming the priority of the physical perspective, while dualism is mistaken in the assumption that two equally fundamental conceptual systems must refer to metaphysically distinct realms of reality. The mind-body problem is ill-conceived because it presents a faulty alternative: either only the physical is fundamental, or we have to accept dualism. According to Schlick, we can accept the fundamentality and irreducibility of the mental without being committed to dualism by taking mental and physical concepts to be equally fundamental.

2.2   The Forgotten Mainstream of Psychophysical Parallelism

Although Schlick's parallelism is at best a fringe position in contemporary philosophy of mind, his parallelism reflected the philosophical mainstream in Germany of the late nineteenth and the early twentieth century. Countless philosophers and scientists endorsed parallelism as a "scientifically minded" alternative to idealism and dualism that also avoided the philosophical problems of materialism.

Materialism arrived late and rather unsuccessfully to German philosophy. Only in the middle of the nineteenth century did a group of biologists and physicians, such as Carl Vogt, Jakob Moleschott, and Ludwig Büchner adopt materialism as a radical alternative to the conservative mainstream of German Idealism and *Naturphilosophie*.[3] The materialist scientists were not particularly interested in the theoretical shortcomings of German Idealism, but understood materialism as a revolutionary movement that combined the rejection of an immaterial soul, god, and free will with a political challenge of the conservative and religious establishment. Despite its public success, academic philosophers largely ignored the materialist movement, or ridiculed it as the vulgar world view of philosophical illiterates. One of the few serious philosophical discussions of materialism was Friedrich Albert Lange's *History of Materialism*, which acknowledged materialism as a methodological doctrine, but rejected its popular metaphysical variants as philosophically naïve. For Lange as well as many other German philosophers of the second half of the nineteenth century, the most obvious shortcoming of the popular materialists was their ignorance with respect to Kant. According to Lange, the popular materialists simply confused scientific entities with the *things-in-themselves*. Consequentially, they had to claim that everything including the mind is scientifically explicable.

Lange was not alone in the assumption that materialism implies unreasonably strong explanatory assumptions. In a highly influential lecture, "On the Limits of our Understanding of Nature," the physiologist Emil du Bois-Reymond presented a generalized argument for the irreducibility of the mental that will sound familiar to contemporary philosophers of mind:

> What conceivable connection is there between certain movements of certain atoms in my brain on one side and on the other the original and undeniable facts that do not allow any further definition: 'I feel pain, feel lust, taste sweetness, smell the scent of roses, hear the sound of an organ, see redness?' (1872, 26).

While the intellectual establishment in Germany unanimously agreed on the rejection of materialism as untenable and philosophically superficial, a large group of philosophers and scientists found the common alternatives of dualism and idealism equally unsatisfying.

Psychophysical parallelism emerged in this context as a scientifically minded but philosophically sophisticated approach to the mind-body problem. The psychophysicist Gustav Fechner deserves credit as a founding figure of this remarkable but almost forgotten episode in the history of the philosophy of mind. Fechner's parallelism dates back to the early 1850s, and was already expressed in his three-volume *Zend-Avesta, or Concerning Matters of Heaven and the Hereafter* (1851). According to Fechner, any successful theory of the mind has to recognize that "the same processes can be perceived as bodily-organic on the one hand or psychologically-mental on

---

[3] Gregory (1977) is an excellent (and the only) monograph on this materialist movement. See also Bayertz et al. (2012) for a helpful anthology.

the other" (1851, 320). Fechner stresses that this account of the mind-body relation does not only imply the rejection of dualism, but also the rejection of a causal interpretation of mind-body interaction: "There are no heterogeneous essences that could interact with each other but one essence that can be perceived from different standpoints. There are not two different, interacting causal chains but one causal chain with one substance that can be traced in two different ways from two different standpoints" (1851, 320).

Fechner's presentation of mental and physical processes as "the same processes" · is clearly incompatible with any substantive dualism. However, at least for the younger Fechner, materialism and idealism were equally flawed in assuming that there is only one fundamental standpoint. In his groundbreaking *Elements of Psychophysics*, Fechner introduced his psychophysical parallelism with the following analogy:

> The solar system seen from the sun appears quite differently than seen from earth. There it is the Copernican, here the Ptolemaic world-system. And for all time it will remain impossible for the same observer to observe both world-systems at the same time, although they both belong inseparably together. Just like the concave and convex side of a circle are two different appearances of the same thing perceived from different standpoints. [...]
> The world is full of examples that show that the same reality will appear in two different ways from two different standpoints and that one cannot substitute one standpoint for another. Who wouldn't admit the ubiquity and necessity of this? Only with respect to the biggest and most convincing example it is not admitted or not thought of. This example is provided by the relationship between the mind and body. (1860, 3).

Fechner's early accounts of psychophysical parallelism mark the beginning of a philosophical mainstream that reached from the 1850s well into the twentieth century and the Weimar Republic. Psychophysical parallelism was almost unchallenged among psychologists and physiologists such as Hermann Ebbinghaus, Edwald Hering, and Wilhelm Wundt. The parallelist doctrine allowed scientists to take the mind and psychological research seriously without having to assume causal powers that would interfere with biological and physiological research. The promise of a sober scientific attitude without the problems of materialism made psychophysical parallelism widely popular and even endorsed by physicists such as Albert Einstein and Niels Bohr (cf. Heidelberger 2004, 177).

In academic philosophy, psychophysical parallelism remained rivaled by different neo-Kantian approaches that leaned towards idealism, but also had influential proponents, including the Berlin-based philosopher, Alois Riehl. Riehl's *Philosophy of Criticism and its Significance for Positive Science* (1876–1887) presented a "realist interpretation" of Kant, and tried to combine this critical realist philosophy with the recent advancements of science (cf. Heidelberger 2006). Riehl's realism is closely connected to his variant of psychophysical parallelism, which he dubbed "identity theory." According to Riehl, the physical and the psychological are two ways of referring to the things in themselves. Riehl's influence on the philosophical reception of psychophysical parallelism is clear, and especially extends to Schlick, who adopted the combination of a parallelist account of mind-body problem with a realist attitude.

Given the widespread endorsement of psychophysical parallelism, it would be wrong to consider Schlick's challenge an obscure and isolated position. On the contrary, psychophysical parallelism was *the* mainstream position in Germany. Michael Heidelberger even speaks of a "peaceful and fruitful rule of psychophysical parallelism in German culture" until the end of the nineteenth century (2004, 246). Of course, psychophysical parallelism was not without critics and in the early twentieth century a growing number of idealist philosophers lamented the proximity of parallelism to materialism. However, the significance of psychophysical parallelism as the dominant, "scientifically minded" philosophy of mind in Germany reached well into the Weimar era, thus placing Schlick's challenge in the rich context of a 70-year old tradition.

2.3 The Ontological Priority of the Physical

The long-lasting philosophical dominance of psychophysical parallelism in Germany raises an obvious question: what happened? How did a mainstream alternative to dualism and materialism, which was endorsed by influential thinkers such as Fechner, Wundt, Schlick, and even Einstein vanish so quickly? It is tempting to explain the sudden disappearance of psychophyscial parallelism through political and institutional factors. As a "scientifically minded" approach to the mind-body problem, psychophysical parallelism was not an attractive theory for Nazi ideologists, and many proponents of psychophysical parallelism either lost their academic positions or were forced into emigration. Furthermore, post-war philosophy of mind became almost exclusively English-speaking, which further reduced the influence of thinkers such as Fechner, Riehl, and Schlick.

There is certainly some truth in this explanation. For example, many of Fechner's and Riehl's works remained untranslated, and even Schlick's *General Theory of Knowledge* didn't appear in English before 1974. However, there are at least two reasons to doubt that these external factors are a sufficient explanation of the disappearance of psychophysical parallelism. First, the disappearance of parallelism from post-war philosophy of mind is matched by an almost equally rapid decline of related philosophical doctrines in America and Great Britain, such as Russell's "neutral monism". Second, the crucial arguments of psychophysical parallelism were present in early post-war philosophy of mind, and can be found in the works of highly influential emigrants, including the Gestalt theorist Wolfgang Koehler (1924), the neuropsychologist Kurt Goldstein (1934, 206–207, cf. Ludwig 2012), and the philosopher Herbert Feigl (1967, cf. Stubenberg 1997). However, these approaches failed to have a recognizable impact on philosophy of mind as it is known today. Feigl's 1958 essay, *The "Mental" and the "Physical"*, is at once the most telling and also the most ironic example, since it is commonly cited as a founding document of contemporary philosophy of mind, despite its advocacy of a largely ignored approach to the mind-body problem.

Feigl does not speak of "psychophysical parallelism," but instead uses labels such as "identity theory," "double knowledge theory," and "double language theory." However, he explicitly acknowledges the historical connections: "My own view is a development in more modern terms of the epistemological outlooks common to Riehl, Schlick, [and] Russell" (1967, 79). Feigl's advocacy of a position that is very close to the tradition of psychophysical parallelism casts doubt on the common presentation of the recent history of philosophy of mind. Usually, the identity theories of Hebert Feigl, Ullin Place, and Jack Smart are presented as a new way of thinking about the mind-body problem. After centuries of misguided dualist and idealist dominance, so the story goes, this group of philosophers formulated the first compelling physicalist theory by assuming the identity of mental and physical states.

Unfortunately this story is misleading in a least two ways. First, the assumption of the identity of mental and physical states was not a novel development. For example, Fechner formulated his psychophysical parallelism as an identity theory in the 1850s, and mind-body identity had been a mainstream position advocated by countless philosophers and scientists. Second, Feigl and Smart subscribed to two very different mind-body theories. While Smart aimed at a materialist identity theory, Feigl insisted on two equally fundamental perspectives. In *The "Mental and the "Physical"*, Feigl writes: "Speaking 'ontologically' for the moment, the identity theory regards sentience (qualities experienced, and in human beings knowledgeable by acquaintance) and other qualities (inexperienced and knowledgeable only by description) as basic reality" (1967, 107). In sharp contrast, Smart argued that there "does seem to be, so far as science is concerned, nothing in the world but increasingly complex arrangements of physical constituents" (1959, 143). And even more emphatically, he wrote: "For on this view there are, in a sense, no sensations. A man is a vast arrangement of physical particles but there are not over and above this, sensations or states of consciousness" (1959, 142).

While Feigl assumes two different but equally fundamental perspectives, Smart argues for exactly one fundamental and ontologically prior level: the physical. I think Smart's comments point toward a more adequate understanding of the recent history of philosophy of mind. The remarkable shift from pre-war to post-war philosophy of mind is commonly misidentified with the emergence of identity assumptions, which were hardly new, but rather had been a mainstream philosophical position since the second half of the nineteenth century. Instead, the assumption of the ontological priority of the physical is arguably the most remarkable feature of post-war philosophy of mind. For Smart and countless other young post-positivist realist philosophers, the physical was not simply one way of referring to reality, but the ontologically fundamental base. Only physics describes what exists in the most fundamental sense and the main task for philosophers of mind is to make sense of the mind in terms of this restricted base of truly fundamental entities. This interpretation of the priority of the physical presents the mind-body problem as a placement problem and clearly separates post-war philosophers of mind from psychophysical parallelism in the tradition of philosophers like Fechner, Riehl, Schlick, and Feigl. If the crucial question is how we can make sense of the mind in a fundamentally physical world, psychophysical parallelism is not even an option as it would require a rejection of the very idea of one metaphysically fundamental way of conceiving reality.

2.4   The Placement Problem in Contemporary Philosophy of Mind

A lot has changed since the publication of Place's and Smart's classical formulations of the identity theory with the majority of contemporary philosophers of mind rejecting type identity and a vocal minority also rejecting physicalism. Despite the diversity of metaphysical positions, there can be little doubt that most philosophers of mind continue to approach the mind-body problem as a placement problem that is created by the assumption of the ontological priority of the physical: the core challenge of philosophy of mind is to find a place for the mind in a fundamentally physical world.

Eliminative and reductive variants of physicalism meet this challenge in the most obvious way by rejecting the existence of entities that are not reducible to a fundamental physical ontology. Furthermore, non-reductive physicalism can be seen as an attempt of making sense of the ontological priority of the physical while avoiding the epistemological commitments of reductionism. The distinction between a metaphysical and an epistemological component in traditional reductive physicalism is a common and well-known topos of non-reductive physicalism (cf. Block 1997). On the one hand, there is the metaphysical claim that only the physical is truly fundamental. On the other hand, there is the epistemological claim that we can explain the mental in terms of the fundamental physical level. Non-reductive physicalism can be understood as the project of showing that we can have the former without the latter. And, of course, there is a huge variety of specifications of this project in terms of constitution, emergence, realization, supervenience, token identity, and so on (cf. Francescotti 2014).

If we look at influential dualist positions in contemporary philosophy of mind such as Chalmers (1996), Jackson (1982), and Kim (2005), we also find discussions of the mind as a placement problem. According to most prominent contemporary dualists, we can stick with the idea of a compact base of fundamental entities and the assumption that almost everything can be understood in terms of a fundamental physical ontology. Phenomenal properties are the only (or at least one of the very few) exceptions and therefore have to be added to the metaphysically fundamental base. For example, Chalmers (2012) defends a scrutability thesis according to which every truth is scrutable from a base PQTI that contains four elements: physical truths (P), phenomenal truths (Q), indexical truths (I), and a negative "that's all" clause (T). In other words, phenomenal and indexical truths constitute the only unsolved placement problems and therefore have to be added to the base of fundamental truths.

Given the framework of placement problems, we only have three credible options in philosophy of mind: We can (1) reduce mental entities to fundamental physical entities, (2) we can eliminate mental entities from our ontologies, or (3) we can extend the base of fundamental entities. Psychophysical parallelism is not a credible option within this framework as parallelists like Schlick reject the very idea of a base of fundamental entities and suggest different but equally fundamental ways of conceptualizing reality.

Even if the idea of the ontological priority of the physical provides a historical explanation of the decline of parallelism in post-war philosophy of mind, it does not provide a justification. On the contrary, the main claim of parallelist philosophers is that we should avoid the assumption of one fundamental characterization of reality, no matter whether it is characterized in materialist/physicalist, dualist, or idealist terms. An exclusion of parallelism through the framework of placement problems therefore begs the question of why we should endorse this framework in the first place.

The main goal of this book is to argue that philosophers of mind should not presuppose a framework of placement problems and that explanatory gaps can be understood in terms of conceptual pluralism. While I consider this pluralism to be broadly in the spirit of a Schlickean parallelism, there are also obvious differences and my discussion will largely rely on contemporary debates in philosophy of science and metaontology. Still, my proposal of conceptual pluralism agrees with at least four core claims of Schlick's *General Theory of Knowledge:* (1) the mind-body problem is ill-conceived; (2) the key to the dissolution of the mind-body problem is the acknowledgment of different but equally fundamental conceptual resources; (3) the result is a theory of mind that should be distinguished from both dualism and materialism; and (4) the acknowledgment of different but equally fundamental conceptual resources is compatible with a robust realism.

References

Bayertz, Kurt, Myriam Gerhard, and Walter Jaeschke, eds. 2012. *Der Materialismus-Streit.* Hamburg: Meiner Verlag.

Block, Ned. 1997. Anti-Reductionism Slaps Back. *Noûs* 31 (s11): 107–32.

du Bois-Reymond, Emil. 1872. *Über die Grenzen des Naturerkennens.* Leipzig: Von Veit & Com.

Braddon-Mitchell, David, and Frank Jackson. 2007. *The Philosophy of Mind and Cognition.* Oxford: Blackwell.

Campbell, Neil. 2005. *A Brief Introduction to the Philosophy of Mind.* Peterborough: Broadview Press.

Chalmers, David. 1996. *The Conscious Mind: In Search of a Fundamental Theory.* Oxford: Oxford University Press.

Chalmers, David. 2012. *Constructing the World.* Oxford: Oxford University Press.

Davidson, Donald. 1970. Mental Events. In *Essays on Actions and Events.* Oxford: Clarendon Press.

Fechner, Gustav. 1851. *Zend Avesta oder über die Dinge des Himmels und des Jenseits.* Leipzig: L. Voss.

Fechner, Gustav. 1860. *Elemente der Psychophysik*. Leipzig: Breitkopf und Härtel.

Feigl, Herbert. 1967. *The Mental and the Physical: The Essay and a Postscript*. Minneapolis: University of Minnesota Press.

Francescotti, Robert. 2014. *Physicalism and the Mind*. Berlin: Springer.

Goldstein, Kurt. 1934. *Der Aufbau des Organismus*. Den Haag: Nijhoff.

Goodman, Nelson. 1978. *Ways of Worldmaking*. Indianapolis: Hackett Publishing.

Gregory, Frederick. 1977. *Scientific Materialism in Nineteenth Century Germany*. Berlin: Springer.

Heidelberger, Michael. 2004. *Nature from Within: Gustav Theodor Fechner and His Psychophysical Worldview*. Pittsburgh: University of Pittsburgh Press.

Heidelberger, Michael. 2006. Kantianism and Realism: Alois Riehl (and Moritz Schlick). In *The Kantian Legacy in Nineteenth-Century Science*, eds. Michael Friedman and Alfred Nordmann, 227–247. Cambridge, Mass.: MIT Press.

Heil, John. 2012. *Philosophy of Mind: A Contemporary Introduction*. New York: Routledge.

Hill, Christopher S., and Brian P. McLaughlin. 1999. There Are Fewer Things in Reality Than Are Dreamt of in Chalmers' Philosophy. *Philosophy and Phenomenological Research* 59: 448–449.

Jackson, Frank. 1982. Epiphenomenal Qualia. *The Philosophical Quarterly* 32: 12.

Kim, Jaegwon. 2005. *Physicalism, or Something Near Enough*. Princeton: Princeton University Press.

Köhler, Wolfgang. 1924. Bemerkungen zum Leib-Seele Problem. *Deutsche Medizinische Wochenschrift* 50: 1269–1270.

Kuhn, Thomas S. 1962. *The Structure of Scientific Revolutions*. Chicago: University of Chicago Press.

Loar, Brian. 1990. Phenomenal States. *Philosophical Perspectives* 4: 81–108.

Ludwig, David. 2012. Language and Human Nature: Kurt Goldstein's Neurolinguistic Foundations of a Holistic Philosophy. *Journal of the History of the Behavioral Sciences* 48 (1): 40–54.

Schlick, Moritz. *General Theory of Knowledge*. Vienna: Springer, 1918/1974.

Schlick, Moritz. Letter to Ernst Cassirer, 1927. Inv. No. 94. Schlick-Papers.

Smart, John JC. 1959. Sensations and Brain Processes. *The Philosophical Review* 68 (2): 141–56.

Stubenberg, Leopold. 1997. Austria Vs. Australia: Two Versions of the Identity Theory. In *Austrian Philosophy Past and Present*, eds. Rudolf Haller, Keith Lehrer, and Johann Christian Marek, 125–46. Boston: Kluwer Academic.

Part II
In Defense of Conceptual Relativity

Chapter 3
Conceptual Relativity in Philosophy

The aim of the following chapters is to develop a general account of conceptual pluralism that is clearly distinguished from a merely epistemological pluralism that has no ontological implications and a strong metaphysical pluralism that argues for exactly one fundamental pluralist ontology. My case for conceptual pluralism will rest on disputes about scientific ontologies and I will argue that we often find a plurality of explanatory interests in science that lead to different but equally fundamental ontologies. Although I will claim that this pluralism of different but equally fundamental ontologies is well-justified in scientific practice, I anticipate the objection that conceptual pluralism has actually highly controversial philosophical implications. Most importantly, conceptual pluralism comes with the negative claim that we should reject the ideal of exactly one fundamental ontology that "carves nature at its joints" and reflects the fundamental structure of reality independently of any contingent conceptual choices. Given that this negative claim contradicts many variants of metaphysical or ontological realism, I will first discuss some of the basic philosophical issues before returning to ontological plurality in the empirical sciences.

While conceptual pluralism may have controversial philosophical implications, it also shares crucial assumptions with many deflationist positions in contemporary metaontology that are skeptical of the ideal of exactly one fundamental ontology (e.g. Putnam 1987; Chalmers 2009; Hirsch 2011). In the following, I will focus on Putnam's metaontological work and argue that large parts of conceptual pluralism can be understood in terms of Putnam's framework of conceptual relativity. Putnam argues that we can describe reality in terms of different conceptual frameworks that imply a plurality of equally legitimate ontologies. While his account of conceptual relativity helps to specify conceptual pluralism, it also illustrates the controversial character of my metaontological claims. In sharp contrast to Putnam's conceptual relativity, many metaphysical and ontological realists insist on the idea of exactly one fundamental account of reality that implies exactly one fundamental ontology.

The most obvious motivation for ontological pluralism is the large diversity of entities that we encounter in reality. Even if we ignore the colorful "bestiary of the

© Springer International Publishing Switzerland 2015
D. Ludwig, *A Pluralist Theory of the Mind*, European Studies
in Philosophy of Science 2, DOI 10.1007/978-3-319-22738-2_3

manifest image" (Dennett 2013) and focus on scientific ontologies, we still face an overwhelming diversity of entities that are postulated in often highly specialized research programs. While this diversity of entities may provide a helpful motivation of pluralism, metaphysicians will point out that there is an important difference between the *de facto* plurality of ontologies that we encounter in scientific practice and a substantive philosophical pluralism that rejects unification efforts that aim at one fundamental ontology.

In formulating a substantive account of conceptual pluralism, it is helpful to start with Putnam account of "conceptual relativity" that he has developed and specified for over 20 years (e.g. 1987, 1988 107–112, 2009 33–52, 2012 29–33). The basic idea of conceptual relativity can be formulated as the positive claim that there are always different but equally fundamental ways of describing reality or as the negative claim that there is not only one fundamental way of describing reality. Furthermore, Putnam combines this epistemic claim about a plurality of equally fundamental descriptions with a metaontological claim about a plurality of equally fundamental ontologies. Consider one of Putnam's earlier examples of conceptual relativity as an illustration of the link between conceptual and ontological issues:

> Suppose I take someone into a room with a chair, a table on which there are a lamp and a notebook and a ballpoint pen, and nothing else, and I ask, 'How many objects are there in this room?' My companion answers, let us suppose, 'Five.' 'What are they?' I ask. 'A chair, a table, a lamp, a notebook, and a ballpoint pen.' 'How about you and me? Aren't we in the room?' My companion might chuckle. 'I didn't think that you meant I was to count people as objects. Alright, then, seven.' 'How about the pages of the notebook?' […]. (1988, 110).

Putnam's conversation illustrates that ordinary language allows different descriptions of the imagined room. In some situations we might be inclined to count people as objects; in other situations we focus only on inanimate things. In some situations, we might count individual pages as objects, whereas in others this may not occur to us. Taking this even further, if we would put one of those pages under a microscope we would again create a very different context, in which we may feel inclined to count other objects such as molecules or atoms. It seems almost trivial that different descriptions can be equally correct. Furthermore, Putnam takes this observation about different descriptions to have straightforward ontological implications: different descriptions imply the existence of different objects and therefore different ontologies. Finally, if we assume that different descriptions should be considered equally correct, we should also accept a plurality of equally correct ontologies.

Few metaphysicians will be impressed by Putnam's example. Even if we assume that ordinary language comes with diverse conceptual resources that imply diverse ordinary ontologies, many metaphysicians will object that philosophical ontology should be clearly distinguished from ordinary ontologies. While it may be an uncontroversial anthropological fact that people use a large variety of ontologies in everyday life,[1] philosophical ontology should aim at a fundamental ontology.

---

[1] One interesting area of research that is concerned with similarities and differences between folk ontologies is ethnobiology. Berlin (1992) and Atran and Medin (2008) provide helpful overviews for research on cross-cultural issues in folk-biological ontologies. See Ludwig (2015) for a detailed discussion.

In other words, Putnam's observation about different ways of describing a room in ordinary language will not be seen as threat for proponents of the ideal of a fundamental account of reality that is uncovered by philosophical and/or scientific research and implies exactly one fundamental ontology.

It is not hard to find influential proponents of the ideal of exactly one fundamental account of reality as it is, for example, expressed in Williams' account of an "absolute conception" of reality (Williams 2011, 153) or Nagel's famous notion of a "view from nowhere" (Nagel 1989, cf. Fine 1998). Furthermore, the idea of exactly one fundamental and absolute description of reality is equally prominent in contemporary metaphysical debates (cf. Blackburn 2010 or Mulder 2012) and nicely expressed in the opening paragraph of Sider's *Writing the Book of the World*: "The world has a distinguished structure, a privileged description. For a representation to be fully successful, truth is not enough; the representation must also use the right concepts, so that its conceptual structure matches reality's structure. There is an objectively correct way to 'write the book of the world'" (Sider 2012, vii).

Although there are important differences between the metaphysical projects of philosophers such as Williams, Nagel, or Sider, they are united in their advocacy of the metaphysical idea of exactly one account of reality that is also taken to imply one fundamental ontology.[2] In the following, I will first introduce the ideal of one fundamental ontology in more detail (Sect. 3.1). Afterwards, I will present Putnam's case for conceptual relativity as an argument against one fundamental ontology (Sect. 3.2) and discuss some more recent deflationist strategies in metaontological debates (Sect. 3.3). As metaontological debates are notorious for ending in contradicting intuitions about the understandability of existence questions, I will turn to philosophy of science and argue for conceptual relativity on the basis of what I have called a "naturalism of scientific practice" (Chap. 4).

3.1 The Idea of a Fundamental Ontology

Ontology is often presented as the philosophical discipline that is concerned with the question "what exists?"[3] However, it is far from obvious that philosophers can provide an interesting answer to this question. Quine famously claimed that the question can be answered in one word: everything. While everyone should agree that everything exists, the answer is obviously not helpful as long as we do not know

---

[2] The relation between "metaphysics" and "ontology" is a complicated issue due to the vagueness of both terms. I will follow a common use of both terms by considering metaphysics to be generally concerned with the structure of reality and ontology to be concerned more specifically with issues of existence. Even if this distinction is often important, Putnam's argument from conceptual relativity clearly combines both issues by rejecting the ideal of exactly one fundamental ontology on the basis of a general rejection of the ideal of one fundamental account of reality.

[3] Most famously Quine (1948, 21); cf. Chalmers (2009, 77). See Guarino et al. (2009) for the wide range of uses of "ontology" in philosophy and science.

the extension of "everything." Drawing up a list of everything that exists would not be helpful either. There are rocks, nuclear power plants, misunderstandings, philosophy departments, art history students, tigers, jazz concerts, and so on. Such a list does not constitute an interesting philosophical insight. How, then, should we understand ontology as a philosophical discipline? One might object that the existence of the listed entities is trivial and that interesting ontological questions arise if we turn to disputed existence claims, but this is to over-simplify the situation. Consider the following example:

(1) Tasmanian tigers are still alive.

Although most biologists agree that the Tasmanian tiger became extinct in the early twentieth century, some cryptozoologists believe that Tasmanian tigers might still be alive. If we take these cryptozoologist claims seriously for the sake of the argument, (1) is a disputed existence claim. Nevertheless, (1) has nothing to do with the project of a philosophical ontology. The question whether Tasmanian tigers still exist is to be answered by zoologists and not philosophers.

Philosophical ontology therefore neither aims at a list of existing entities nor at a discussion of all disputed existence claims. We get a step closer to understanding to philosophical ontologies when we realize that ontologists typically discuss very general existence claims, leaving specific existence claims to the empirical sciences and our everyday discussions. For example, ontologists do not discuss questions such as:

(2) Is there a table in the room?
(3) Are there more than 10.000 Tasmanian devils?
(4) Are there trees in Greenland?

Rather, ontologists typically discuss questions such as:

(5) Are there macroscopic objects such as tables, Tasmanian devils or trees?

The distinction between specific and general existence questions brings us closer to an understanding of philosophical ontology. Ontologists are usually not interested in specific existence questions such as (2) – (4), but try instead to describe the general structure of reality. The distinction can also be applied to other pairs of examples:

(6) Are there more than four prime numbers smaller than 10?
(7) Are there numbers?
(8) Are there any interesting jazz concerts in Berlin next Tuesday night?
(9) Are there events?

Again, the distinction between specific and general questions seems to explain the distinction between ontologically innocent questions like (6) and (8), and supposedly deep ontologically questions like (7) and (9). However, we immediately run into a new complication, as most of the general existence claims seem to have obvious answers. Consider the question whether there are macroscopic objects such as tables, Tasmanian devils or trees. Answers to specific existence questions seem to imply answers to this general existence question:

There are two tables in the room.
Tables are macroscopic objects.
∴ There are macroscopic objects.

The same problem occurs in the case of (6) and (7), as well as with (8) and (9): there are four prime numbers smaller than 10. Prime numbers are numbers. Therefore, there are numbers. There are interesting jazz concerts in Berlin on Tuesday night. Jazz concerts are events. Therefore, there are events.

There are at least two reasons to be suspicious of these arguments. First, they imply that ontological questions have truly trivial answers, and it becomes hard to understand how anyone could ever have considered ontology an interesting philosophical project. Furthermore, many contemporary ontologists reject the allegedly trivial conclusions. For example, many philosophers have argued that macroscopic objects such as tables do not really exist.[4] While these philosophers surely utter sentences such as (10), they still believe that (11) is true.

(10) There are two tables in the room
(11) There are no macroscopic objects such as tables

How is this possible? It is hard to believe they missed the fact that (10) seems to contradict (11). One strategy to solve the *prima facie* contradiction between (10) and (11) is suggested by Rudolf Carnap's (1950) famous distinction between internal and external questions. According to Carnap, internal existence questions are posed within specific conceptual frameworks and often have uncontroversial answers. For example, given the conceptual framework of ordinary English, there are two tables in the room and there are jazz concerts in Berlin on Tuesday nights. Given the conceptual framework of mathematics, there are four prime numbers smaller than 10, and given conceptual framework of zoology, there are at least 10,000 Tasmanian devils.

Many contemporary ontologists, however, will insist that they are not interested in internal existence questions, but want to know what exists independently from our conceptual frameworks. In other words, they are interested in external existence questions. The distinction between internal and external questions solves the *prima facie* contradiction between (10) and (11):

(10)' There are$_{\text{internal}}$ two tables in the room

contradicts

(11)' There are$_{\text{internal}}$ no macroscopic objects such as tables

but it does not contradict

(11)'' There are$_{\text{external}}$ no macroscopic objects such as tables

---

[4] Sellars (1963, 62–75); Unger (1979); van Inwagen (1990, 81–97); Merricks (2001) are influential positions that challenge ordinary objects in one way or another. Unger (2005), however, has abandoned his eliminativism about macrophysical objects.

Carnap's distinction between internal and external questions carries a lot of historical baggage[5] and was introduced with the intention of rejecting external questions as having no truth value. However, we can reformulate the basic idea without Carnap's terminology by differentiating between ordinary and fundamental existence questions.[6] Ordinary existence questions are similar to Carnap's internal questions in the sense that they are asked in contexts in which we usually have well-known rules of how to answer them. While we know how to answer questions about ordinary ontologies, many metaphysicians assume that philosophers should be concerned with the additional question what entities exist in the most fundamental sense. While ordinary existence questions may lead to a plurality of ordinary ontologies, philosophical ontology should aim at one truly fundamental ontology that reflects the basic structure of reality.

The distinction between ordinary and fundamental existence questions solves the *prima facie* contradiction between (10) and (11). Philosophers who reject the existence of macroscopic objects can admit that there are everyday contexts in which it is correct to say that there are two tables in the room. However, they will insist that our ordinary claims do not have any ontological implications and do not answer the question of whether macroscopic objects like tables exist in a fundamental sense. It may be the case that there are no macroscopic objects, but only elementary particles that can be arranged in different ways. Maybe there are no tables, but only "table-wise arranged elementary particles."[7] In such a situation, it might still be correct to say that there are two tables in the room in everyday contexts, but tables do not *really* exist.

The distinction between ordinary and fundamental existence questions is based on the assumption that there is a fundamental ontological account, which allows us to go beyond our ordinary existence claims. Putnam illustrates this point by distinguishing between an ontology that is concerned with all of our ordinary and scientific existence claims and the ideal one fundamental Ontology (with a capital "O") that transcends the plurality of ontologies that we encounter in everyday contexts and scientific practice (Putnam 2009, 21). The ambitious ideal of Ontology clearly challenges conceptual pluralism: even if we find a large diversity of ontologies in everyday contexts and scientific practice, the aim of philosophical Ontology is to step beyond them and to discuss what entities exist *in the most fundamental sense*. At the same time, one may wonder why we should accept this ideal of Ontology

---

[5] Eklund (2013) argues that Carnap-references in contemporary metaphysics are largely based on a misunderstanding of this historical debate.

[6] Chalmers' (2009, 80–85) proposes an analogous distinction between ordinary and ontological existence questions. "Ontological existence questions" in the sense of Chalmers are exclusively concerned with *fundamental* ontological truths and therefore correspond to my "fundamental existence questions".

[7] This strategy is spelled out by Van Inwagen (1990, 98–114). See Thomasson (2007) for a critical discussion.

with a capital "O" and Putnam's arguments for conceptual relativity attempt to show that we should instead endorse a liberal ontological pluralism.[8]

3.2 Putnam's Case for Conceptual Relativity

Analytic metaphysicians often propose positions that radically contradict ordinary as well as scientific ontologies. For example, Peter van Inwagen's highly influential *Material Objects* rejects the existence of most ordinary and scientific objects but accepts the existence of organisms. According to van Inwagen's organicism, objects compose a new object if and only if they constitute a life (1990, 90). This means that there are no tables, because the "table-wise arranged elementary particles" do not constitute a life and no genes because "gene-wise arranged elementary particles" do not constitute a life. However, there are Tasmanian devils because the "Tasmanian-devil-wise arranged elementary particles" constitute a life. Other metaphysical positions are even more radical as they reject the existence of all composed objects or accept the existence of all artificially composed objects such as an object that is composed by China and your morning cereal.

Not everyone is able to take these revisionary ontologies seriously. Dennett (2013), for example, warns us of "analytic metaphysics and other dubious battles in philosophy" (101) and recommends the perspective of a "diplomatic anthropologist, not the metaphysician intent on limning the ultimate structure of reality." Hirsch is even more polemical in summarizing van Inwagen's organicism and his own doubts: "'All things considered, I am tentatively inclined to be ontologically committed to apple trees but not to apples.' The challenge [...] posed by this kind of formulation is how to keep a polite straight face while listing to it." (2002, 67) Poking fun at contemporary ontology, however, will not convince philosophers who clearly take revisionary ontologies seriously.

In order to develop a more substantive criticism of the ideal of one fundamental ontology and its revisionist consequences, it is helpful to have a closer look at Putnam's case for conceptual relativity. In his 1987 article, "Truth and Convention," Putnam argues that conceptual relativity not only affects ordinary language, but also allegedly fundamental ontological existence questions. Consider a universe with three individuals, $x1$, $x2$, and $x3$, and the question of how many objects exist. It may seem obvious that there are three objects: $x1$, $x2$, and $x3$. However, it is also possible

---

[8] Putnam's criticism of Ontology with a capital "O" occasionally leads him to an anti-ontological rhetoric as illustrated by the title of his *Ethics without Ontology* (2004 cf. Pihlström 2006 and Copp 2006). Dale (2008) extends this strategy to debates about mind and cognition by suggesting a "cognitive science without ontology". However, I do not assume that Putnam's or Dale's proposals are incompatible with my presentation of ontological pluralism. Instead, the difference is largely rhetorical: while I talk about "ontology" in a metaphysically shallow sense that is ubiquitous in scientific practice, Putnam and Dale reject "Ontology" as a philosophical project that aims at exactly one fundamental account of what exists.

to claim that the individuals compose new objects such as $(x1 + x2)$ or $(x1 + x2 + x3)$. For example, Putnam introduces a "Polish logician," who believes that for every two individuals, there is an object which is their sum. According to the Polish logician, not only $x1$, $x2$, and $x3$, but also their sums $(x1 + x2)$, $(x2 + x3)$, $(x1 + x3)$, and $(x1 + x2 + x3)$ exist. We are then confronted with different answers to the question how many objects exist:

(12) There are exactly three objects
(13) There are exactly seven objects

Putnam takes this example to show that there is not one fundamental ontology in a universe with three individuals. It would be pointless to discuss whether (12) or (13) are fundamentally true, because the answer does not depend on facts external to us, but on our conceptual decisions. If we choose an ontology that only accepts individuals, (12) is the correct answer. If we choose an ontology that also accepts sums of individuals, (13) is the correct answer. However, we have to make an ontological *decision*, and there is no reasonable question of how many objects really or fundamentally exist in Putnam's universe.

Putnam's argument from conceptual relativity is designed as an attack on the very idea of one fundamental ontology. If ontology aims at an absolute inventory of the universe that is independent of any conceptual choices, then conceptual relativity shows that there is something wrong with the entire project. However, many analytic metaphysicians will disagree with Putnam's diagnosis and insist that there is only one fundamentally correct answer to the question of how many objects exist in Putnam's universe.

The most common answers in contemporary ontology are the most radical ones: "mereological nihilists" argue that individuals never compose new objects and therefore agree with (12). "Mereological universalists" assume that individuals always compose new objects and agree with (13).[9] The controversy is not limited to obscure philosophical examples, but extends to the real world. Nihilists claim that composed objects such as tables, plants, rocks, or genes do not really exist. The ontological claims of universalists are equally bizarre as they claim that every object composes a new object with every other object. For example, there is even an object composed of Immanuel Kant's grave, The Rolling Stones, and Alpha Centauri.

How is it possible that nihilists and universalists disagree over (12) and (13), while Putnam argues that both sentences are equally correct and that it would be obviously pointless to argue whether (12) or (13) are really true? According to nihilists and universalists, their debate is only superficially similar to debates about ordinary ontologies on issues such as whether we should count a book or each of its pages as an object. However, this superficial similarity misleads philosophers like Putnam to "dissolve" the ontological debate by differentiating between different conceptual frameworks. Consider the following reformulation that makes the assumption of different conceptual frameworks explicit:

---

[9] Dorr and Rosen (2002) provide a helpful overview. See Tallant (2014) for a (critical) discussion of nihilism and Van Cleve (2008) for a defense of universalism.

(12)' There are$_{\text{nihilism}}$ exactly three objects
(13)' There are$_{\text{universalism}}$ exactly seven objects

The reformulation dissolves the *prima facie* contradiction. Nihilists and universalists talk about the existence of objects in different ways and thus can be both correct. However, analytic metaphysicians such as nihilists and universalists will insist that they are not interested in (12)' and (13)', but the truth values of the following claims:

(12)" There are$_{\text{absolute}}$ exactly three objects
(13)" There are$_{\text{absolute}}$ exactly seven objects

(12)" and (13)" actually contradict each other; only one of them can be true. This move towards an absolute quantifier explains how proponents of the ideal of one fundamental ontology can respond to Putnam's challenge. They can say something along the following lines: "if we are concerned with fundamental (instead of ordinary) existence questions, there is no conceptual relativity. If (12) and (13) are taken to be ontological claims about the absolute number of objects, then they obviously contradict each other. Putnam is simply wrong in claiming that it is pointless to discuss how many objects exist in a universe with three individuals. If we take the ideal of one fundamental ontology seriously, then the debate between nihilists and universalists is not pointless at all."

3.3   Understandability and the Epistemic Challenge

Putnam does not take the idea of exactly one fundamental ontology (or Ontology with a capital "O") seriously and there are indeed at least two reasons to be skeptical about claims such as (12)" and (13)". First, one can doubt that we actually understand what they mean. Second, one can argue that we would never be able to figure out whether (12)" or (13)" is true.

Understandability Consider real life applications of nihilism and universalism. Nihilists claim that macroscopic objects such as tables do not *really* exist. But what does it mean that tables do not *really* exist? As Chalmers puts it: "one might question whether we really have a grip on what it would be for a table to 'really exist' versus what it would be for a table to fail to exist" (2009, 103). Universalists face the same problem: what does it mean that Immanuel Kant's grave, The Rolling Stones, and Alpha Centauri *really* compose a new object? Do we understand the difference between them composing an object and them not composing an object? It is perfectly legitimate to ask ontologists for an explanation of claims such as "Tables do not really exist," or "There really is an object composed of Immanuel Kant's grave, The Rolling Stones, and Alpha Centauri." However, ontologists do not offer such an explanation and therefore leave us in the dark if we do not already understand the idea of an absolute ontology.

The worry that we do not understand the ontologists' claims can be further motivated. One of the reasons why it is so hard to understand what nihilists and universalists are disagreeing about is that their disagreement appears inferentially isolated in the sense that it is not connected to any other well-understood disagreements. In fact, it appears that both parties can agree about *all* other relevant facts. In the case of Putnam's universe with three individuals, nihilists and universalists agree on the number of individuals, their spatial positions, their status as individuals without proper parts, and so on. In the case of real world examples, they can also agree on all relevant facts including when it is correct to utter sentences such as "There are two tables in the room". Given that the parties can agree on virtually everything, it becomes increasingly unclear what they are disagreeing about.

The Epistemic Challenge The worry that we do not understand the ontologists' ideal of exactly one fundamental answer to the question what entities exist is closely related to another common epistemological worry. Given the inferential isolation of the debate between nihilists and universalists, it becomes unclear how their debate could ever be resolved.[10] We cannot resolve it empirically by conducting an experiment or by observation. In the case of Putnam's universe with three individuals, we can count the individuals, but this will not help as both parties agree that there are exactly three individuals. And, of course, we cannot count the objects without presupposing precisely what is to be shown: the priority of one ontology.

We cannot resolve the debate conceptually either since there is no reason to believe that nihilism or universalism are formally or analytically inconsistent. Of course, one could reject nihilism with the argument that the existence of composite objects is implied by ordinary language. However, a nihilist can respond in two ways. First, she can deny that ordinary language includes any ontological commitments and insist that a nihilist can accept ordinary language sentences as correct (van Inwagen 1990; Horgan and Potrč 2006). For example, a nihilist can claim that there are many possible ontological interpretations of ordinary language, and she will translate a sentence such as "There are tables" into "There are elementary particles arranged table-wise." Second, even if ordinary language were committed to the existence of composite objects, a nihilist could still argue that ordinary language is simply wrong.

Finally, pragmatic considerations will also not resolve the debate. Nihilists and universalists might disagree on pragmatic issues and each claim that their own conceptual proposal is pragmatically superior, but this is obviously not the core issue. Pragmatic considerations are too pluralism-friendly to satisfy proponents of the ideal of one fundamental ontology. For example, ordinary ontologies that contradict both nihilism and universalism are clearly often useful but both nihilists and universalists still want to claim that they are false.

In a recent article, Kriegel (2013) has formulated a general "epistemological challenge of revisionary metaphysics" that is based on these kinds of considerations.

---

[10] Compare Rosen and Dorr (2002) and Bennett (2009) who raise these issues but do not consider them conclusive arguments against the ideal of one fundamental ontology.

According to Kriegel, strong metaphysical claims could be supported by considerations regarding (a) empirical adequacy, (b) intuitions, or (c) "super-empirical" values such as simplicity. However, there is no evidence that strong metaphysical claims can be evaluated on the basis of these considerations and therefore no reason to believe that we can possibly know their truth values. Finally: if we cannot find any evidence for or against the truth of strong metaphysical claims, we have a very good reason to suspect that there is something deeply wrong with them. Hence, the "epistemological challenge of revisionary metaphysics".

The two objections that I have sketched in the previous paragraphs are well-known starting points for debates about the feasibility of ontological projects. Whenever a philosopher presents claims about fundamental ontological truths, she can expect two challenges (1) What does "fundamental ontology" mean? (2) How are we supposed to know about this allegedly fundamental ontology?

Unfortunately, both challenges have a track-record of leaving metaontological debates in a stalemate. While skeptics often assume that their examples illustrate how flawed the ideal of one fundamental ontology is, passionate ontologists will reject the objection of understandability as an unfair intuition pump. Sider, for example, warns us of a "magical grasp picture of understanding" and instead suggests that we can understand metaphysical terms such as "fundamental structure" by clarifying their role in our thinking (Sider 2012, 9). Although Sider is certainly right in warning us of a "magical grasp" picture of understanding, his alternative is unlikely to solve the issue of conflicting intuitions about the understandability of claims in analytic metaphysics. Deflationary intuitions about ontology and metaphysics are often based on the assumption that we do not understand the role of concepts such as "fundamental ontology" or "fundamental structure" in our thinking because they are inferentially disconnected from all other questions we actually have a grip on. Passionate metaphysicians, however, see a thick inferential network of metaphysical issues that leaves little doubt about our understanding of the role of metaphysical terms such as "fundamental structure" in our thinking.

While the issue of understandability has the tendency of reinforcing intuitions of both proponents and critics of analytic metaphysics,[11] the epistemological challenge fares slightly better: even enthusiastic metaphysicians may admit that the "epistemological underpinnings [of metaphysics] are disconcertingly underdeveloped" (Kriegel 2013) and that there is a real challenge that needs to be addressed. However, the acknowledgment of a challenge does not imply that the challenge cannot be met and even a brief look at contemporary debates about issues such as nihilism and universalism provides plenty evidence that many philosophers

---

[11] One may also worry that philosophical intuitions about the understandability or non-understandability of metaphysical issues are easy prey for the "negative program" in experimental philosophy (Alexander et al. 2010; cf. Thomasson 2012) that challenges the use of intuitions in philosophical arguments. If there are no good arguments of *why* we should consider certain metaphysical issues understandable or not understandable, the variability of intuitions provides a further reason to expect a stalemate in the discussion of the ideal of one fundamental ontology.

think that they can actually provide arguments for or against the truth of their preferred ontological theory.

Furthermore, even if the epistemological challenge were successful in showing that we cannot know the truth-values of many metaphysical claims, proponents of the ideal of a fundamental ontology could still bite the bullet and accept that we will never know what fundamentally exists. Although I agree with Kriegel's assessment that this epistemological skepticism is "hard to swallow" (Kriegel 2013), neither Kriegel's nor my intuition is very relevant if it is not shared by proponents of strong metaphysical and ontological programs. Furthermore, there can be little doubt that many enthusiastic metaphysicans will stick with their ideal of exactly one fundamental ontology no matter how bad the epistemological situation turns out to be, because they are convinced that there is no viable alternative.

The rejection of the ideal of one fundamental ontology does not even seem be an option for many of its proponents because they assume that any alternative such "conceptual relativity" will ultimately lead to an unstable and entirely unacceptable relativist and anti-realist picture. Sider expresses this worry by not only warning us of "Goodmania" (2012, 186) as the only alternative to joint carving but also declaring that the failure of his strong metaphysical program would mean that the "postmodernist forces of darkness have won" (2012, 65).

References

Alexander, Joshua, Ron Mallon, and Jonathan M. Weinberg. 2010. Accentuate the Negative. *Review of Philosophy and Psychology* 1 (2): 297–314.

Atran, Scott, and Douglas L. Medin. 2008. *The Native Mind and the Cultural Construction of Nature*. Cambridge, Mass.: MIT Press.

Bennett, Karen. 2009. Composition, Colocation and Metaontology. In *Metametaphysics: New Essays on the Foundations of Ontology,* eds. David John Chalmers, David Manley and Ryan Wasserman, 38–75. Oxford: Oxford University Press.

Berlin, Brent. 1992. *Ethnobiological Classification*. Princeton: Princeton University Press.

Blackburn, Simon. 2010. The Absolute Conception: Putnam Vs Williams. *Practical Tortoise Raising: and other Philosophical Essays,* Simon Blackburn. Oxford: Oxford University Press.

Carnap, Rudolf. 1950. Empiricism, Semantics, and Ontology. *Revue Internationale de Philosophie* 4: 20–40.

Chalmers, David. 2009. Ontological Anti-Realism. In *Metametaphysics: New Essays on the Foundations of Ontology,* eds. David John Chalmers, David Manley and Ryan Wasserman, 77–129, Oxford: Oxford University Press.

Van Cleve, James. 2008. The Moon and Sixpence: A Defense of Mereological Universalism. In *Contemporary debates in metaphysics,* eds. Ted Sider, John Hawthorne and Dean W Zimmerman, 321–366. Malden, MA: Blackwell Publishing.

Copp, David. 2006. The Ontology of Putnam's Ethics without Ontology. *Contemporary Pragmatism* 3 (2): 39–53.

Dale, Rick. 2008. The Possibility of a Pluralist Cognitive Science. *Journal of Experimental and Theoretical Artificial Intelligence* 20(3): 155–179.

Dennett, Daniel C. 2013. Kinds of Things. Bestiary of the Manifest Image. In *Scientific Metaphysics*, eds. Don Ross, James Ladyman, and Harold Kincaid, 96–107. Oxford: Oxford University Press.

Dorr, Cian, and Gideon Rosen. 2002. Composition as Fiction. In *Blackwell Guide to Metaphysics*, ed. Richard M Gale, 151–74. Oxford: Blackwell.

Eklund, Matti. (2013). "Carnap's Metaontology," *Noûs*, *47*(2), 229–249.

Fine, Arthur. 1998. The Viewpoint of No-One in Particular. In *Proceedings and Addresses of the American Philosophical Association*, 7–20.

Guarino, Nicola, Daniel Oberle, and Steffen Staab. 2009. What is an Ontology?. *Handbook on ontologies*, eds. Steffen Staab and Rudi Studer, 1–17, Berlin: Springer Berlin Heidelberg.

Hirsch, Eli. 2002. Quantifier Variance and Realism. *Philosophical Issues* 12 (1): 51–73.

Hirsch, Eli. 2011. *Quantifier Variance and Realism: Essays in Metaontology*. Oxford: Oxford University Press.

Horgan, Terry, and Matjaž Potrč. 2006. Abundant Truth in an Austere World. *Truth and Realism: New Essays*, eds. Michael Lynch & Patrick Greenough, 137–67. Oxford: Oxford University Press.

Van Inwagen, Peter. 1990. *Material Beings*. Cornell: Cornell University Press.

Kriegel, Uriah. 2013. The Epistemological Challenge of Revisionary Metaphysics. *Philosophers Imprint* 13 (12): 1–30.

Ludwig, David. 2015. Indigenous and Scientific Kinds. *The British Journal for the Philosophy of Science*, first published online,.

Merricks, Trenton. 2001. *Objects and Persons*. Oxford: Oxford University Press.

Mulder, Jesse M. 2012. What Generates the Realism/Anti-Realism Dichotomy. *Philosophica* 84: 49–80.

Nagel, Thomas. 1989. *The View from Nowhere*. Oxford: Oxford University Press.

Pihlström, Sami. 2006. Putnam's Conception of Ontology. *Contemporary Pragmatism* 3 (2): 1–13.

Putnam, Hilary. 1987. Truth and Convention: On Davidson's Refutation of Conceptual Relativism. *Dialectica* 41 (1-2): 69–77.

Putnam, Hilary. 1988. *Representation and Reality*. Cambridge: Cambridge University Press.

Putnam, Hilary. 2009. *Ethics Without Ontology*. Harvard: Harvard University Press.

Putnam, Hilary. 2012. From Quantum Mechanics to Ethics and Back Again *Reading Putnam,* eds. Peter Clark and Bob Hale, 19–36. New York: Routledge.

Quine, Willard Van Orman. 1948. On What There Is. *Review of Metaphysics* 2 (5): 21—36.

Sellars, Wilfrid. 1963. Philosophy and the scientific image of man. *Science, perception and reality* 2: 35–78.

Tallant, Jonathan. 2014. Against Mereological Nihilism. *Synthese* 191 (7): 1511–1527.

Thomasson, Amie. 2007. *Ordinary Objects*. Oxford: Oxford University Press.

Thomasson, Amie. 2012. Experimental Philosophy and the Methods of Ontology. *The Monist* 95 (2): 175–99.

Unger, Peter. 1979. There are no ordinary things. *Synthese* 41 (2): 117–154.

Unger, Peter. 2005. *All the Power in the World*. Oxford: Oxford University Press.

Sider, Ted. 2012. *Writing the Book of the World*. Oxford: Oxford University Press.

Williams, Bernard. 2011. *Ethics and the Limits of Philosophy*. Cambridge, Mass.: Harvard University Press.

Chapter 4
Conceptual Relativity in Science

The aim of the following sections is to develop an account of conceptual relativity in scientific practice. Pluralist interpretations of scientific ontologies are common in many areas of the life sciences from microbiology to psychiatry (e.g. Barker and Velasco 2013; Bapteste and Boucher 2009; Kaplan and Winther 2014; Kitcher 2008; Leonelli 2013; Longino 2013; Winther 2011; Zachar 2002). Although pluralist accounts of scientific ontologies obviously do not agree on all philosophical issues, they typically share the starting point of the diversity and contingency of explanatory interests in science. In a second step, it is argued that scientists with different explanatory interests often find different entities meaningful and therefore opt for different ontologies.[1] Finally, many pluralist philosophers insist that this plurality of ontologies is an irreducible aspect of scientific practice and not just a temporary reflection of the current state of research. Ontological unification across all research projects of a discipline such as microbiology or psychiatry is neither a realistic nor a desirable goal. Of course, there is nothing wrong with unification per se and a unified ontology can be a reasonable goal in an interdisciplinary research context. At the same time, there is plenty evidence that the ontological needs of researchers do not always converge and that global ontological unification is not a helpful ideal in scientific practice.

The aim of the following sections is twofold. On the one hand, I will defend a pluralist analysis of scientific ontologies by looking at one prominent case study (4.1 species) and two less common examples of conceptual relativity in science (4.2 cognition and 4.3 intelligence). On the other hand, I will argue that philosophers can learn both a positive and a negative lesson from this plurality of scientific ontologies. The positive lesson is that conceptual relativity is not nearly as shocking as many analytic metaphysicians think it is. The observation that scientists often use a plurality of conceptual frameworks that imply different but equally correct answers

---

[1] Although I will follow debates in philosophy of science, related arguments about explanatory interests are also found in contemporary metaphysics – see, for example, Irmak's (2014) case for the purpose-relativity of ontology.

© Springer International Publishing Switzerland 2015
D. Ludwig, *A Pluralist Theory of the Mind*, European Studies
in Philosophy of Science 2, DOI 10.1007/978-3-319-22738-2_4

to the question what entities exist does not open the gate for Sider's "postmodern forces of darkness" (2012, 65) as it does not challenge the broadly realist claim that the sciences are concerned with a reality that is independent of our conceptualizations. The positive lesson of this chapter can be seen as an offer to philosophers who insist on the ideal of one fundamental ontology because they are worried that anything less will inevitably lead to downward spiral of relativism and anti-realism.

While my discussion of case studies from the empirical sciences has a positive side, it also entails a negative challenge of the ideal of exactly one fundamental ontology (cf. 4.6). If different but equally fundamental ontologies are common and unproblematic in scientific practice, philosophers have to provide very good reasons to stick with the ideal of exactly one fundamental ontology. The negative challenge is especially pressing in the context of a methodological approach that I have described as "naturalism of scientific practice." Given the assumption that philosophers are not in the epistemic position to step behind the empirical sciences, it becomes hard to see how the ideal of one fundamental ontology is viable in the light of ontological plurality in scientific practice. In other words: If our best understanding of scientific practice suggests a plurality of equally fundamental ontologies, philosophers better be prepared to follow.

4.1 Species

The overall goal of this chapter is to argue for conceptual relativity in science and for the idea of a plurality of different but equally fundamental scientific ontologies. Some of the most obvious candidates for conceptual relativity in the empirical sciences come from the vast and diverse literature on "the edges and boundaries of biological objects" (Haber and Odenbaugh 2009). For example, different biological taxonomies imply the existence of different biological entities and in this sense imply different biological ontologies. In this section, I will focus on the best known taxonomic issue in philosophy of biology: the species debate.

Species membership has been traditionally determined along morphological criteria: two organisms belong to the same species, if they share sufficient morphological properties.[2] Of course, not every morphological property is relevant for traditional

---

[2] In discussing species as kinds, I am ignoring a common debate about the ontological status of species. The question whether we should think of species as kinds or individuals has, as Rieppel (2013, 166) points out, "created an industry from which resulted a flow of publications that shows no signs of slowing down." I do not have any objections against accounts of species as individuals. Instead, I assume that there are two reasons why the aims of this chapter are largely independent from the kinds vs. individuals issue. (1) A discussion of species as kinds does not exclude a discussion of species as individuals. For example, Kitcher (1987, 187) argues that both ontologies are equivalent and that we can simply choose our preferred ontological framework: "Fans of mereology will prefer mereological reconstructions and friends of set theory will opt for set-theoretic analyses." More recently, both Reydon (2005, 2009) and Rieppel (2013) have argued that accounts of species as kinds or as individuals are not equivalent but still both correct. For example, Reydon

species concepts. For example, members of the same species do not all have the same size or weight. Furthermore, members of different species can have the same size or weight. An adequate characterization of the morphological species concept therefore has to identify crucial or essential morphological properties that are shared by members of the same species. In the light of modern population biology, however, it has become a truism that an appeal to essential properties is no longer viable. Given intraspecific variation and the dynamic historical character of species, we cannot hope to find a fixed set of essential properties that separate species from each other.

If there are no essential morphological properties, how then can biologists distinguish between different species? One of the best known proposals is Ernst Mayr's so called *biological species concept*. According to Mayr, species are "groups of interbreeding natural populations that are reproductively isolated from other such groups" (Mayr 1973, 21). Mayr's biological species concept suggests the following criterion: individuals belong to the same species iff they are able to produce fertile offspring of both sexes. This proposal offers an elegant solution for the problem of a diverse and changing sets of morphological properties. Consider the genus *Equus*, which includes different species such as the imperial zebra, the wild horse, the African wild Ass, and the Asian wild ass. One problem with morphological species concepts is that they do not clarify which morphological properties we should consider essential for distinguishing between species. For example, we could introduce one general species "ass" or we could consider the African wild ass and the Asian wild ass to be two distinct species. We could also take the different variants of the Asian wild ass (Mongolian wild ass, Turkmenian kulan, Persian onager, Indian wild ass, and Syrian wild ass) to be distinct species. The problem with the morphological species concept is that it seems rather arbitrary where we draw the line. From a morphological perspective, there is no clear answer to the question of whether there is just one ass species, two ass species (African Wild ass and Wild asian ass), or more than two ass species (Mongolian wild ass, Turkmenian kulan, Persian onager, and so on).

While morphological criteria create confusion, the biological species concept offers a simple solution: populations belong to the same species if and only if they can produce fertile offspring of both sexes. Notice here that offspring is not enough: *fertile* offspring *of both sexes* is necessary. If the offspring is not fertile, populations will remain reproductively isolated from each other and there will not be an appropriate gene exchange between them. In the case of *Equus* it is common knowledge that not every offspring is fertile. A mule is the offspring of a male donkey and a

---

argues that "species" is simply ambiguous and can refer to ontologically non-equivalent kinds and individuals. (2) Even if it were true that species are individuals and not kinds, I would only have to slightly change the presentation of my case for conceptual relativity. Indeed, I have introduced conceptual relativity as a claim about scientific ontologies and not as a claim exclusively about scientific kinds. If species were individuals, the species debate would still illustrate conceptual relativity in terms of the availability of different accounts of biological individuals.

female horse. However, all male mules and most female mules are infertile and therefore donkeys and horses belong to different species.

Even if the biological species concept offers an elegant proposal, it is not without problems. Most obviously, the biological species concept cannot be applied to asexual species unless one is willing to accept that asexual individuals do not belong to species at all. Even in the case of species that rely on sexual reproduction, the biological species concept can lead to undesirable results. For example, hybridization is quite common among plants even in cases where we want to distinguish between different species.

One way of dealing with these problems is to endorse van Valen's *ecological species concept*: "A species is a lineage (or a closely related set of lineages) which occupies an adaptive zone minimally different from that of any other lineage in its range and which evolves separately from all lineages outside its range" (1976, 233). Van Valen's proposal offers two necessary and jointly sufficient conditions: in order to belong to the same species, individuals must share an "adaptive zone" (or "ecological niche") and they must be part of the same lineage. The first condition reflects the central idea that natural selection within an adaptive zone is the primary force in preserving species. The second condition clarifies that the ecological species concept recognizes species as genuinely historic kinds with common ancestors.

Compared to the biological species concept, the ecological species concept has the advantage of being able to distinguish between different asexual species: asexual individuals belong to the same species if they are subject to the same set of adaptive forces and they belong to different species if they are subject to different sets of adaptive forces. Furthermore, the ecological species concept is especially attractive to botanists as it can distinguish between different species even in the case of hybridization.

However, the ecological species concept is not the only alternative to Mayr's biological species concept. Another proposal that has gained considerable attention is the so-called *phylogenetic species concept*, which defines species as "the smallest diagnosable cluster of individual organisms within which there is a parental pattern of ancestry and descent" (Cracraft 1983, 170). For many biologists, this proposal is attractive because of its closeness to cladistic research practice and its fine-grained species ontology. However, not every biologist is happy with this proposal and its greatly increased number of species (cf. Agapow et al. 2014). First, the phylogenetic species concept is not the only evolutionarily sensitive proposal. On the contrary, many species concepts consider a common ancestry a necessary condition of species membership, as we have seen in the case of the ecological species concept. Second, the fine-grained species ontologies are not considered a benefit by all biologists. In his scathing book review "Cladistics in Wonderlaned", Avise (2000, 1828) accuses proponents of phylogenetic species concepts of creating a "world of sense and nonsense often turned on its head, of erudite jabberwocky, of impeccably logical illogic, of surreal reality." According to Avise, phylogenetic species concepts inflate the number of species at great cost and little benefit and he ironically remarks that we "may look forward to a twenty-first century in which the rate of species origin (via fixation of genetic variants) may outpace the rate at which currently recognized taxonomic species are driven to extinction" (2000, 1828). Zachos et al. (2013) echo these worries in a more moderate tone and argue that cladistic distinctions between mammalian species are "taxonomic artifacts" that are not only based on inconclusive data sets but also put "an unnecessary burden on the conservation of biodiversity" (2013, 1).

The list of species concepts could be easily extended. For example, Kevin de Queiroz distinguishes between 12 species concepts with different strengths and weaknesses (de Queiroz 1998). However, the goal of this section is not an extensive review of species concepts but an applied discussion of conceptual relativity in scientific practice. And even if we limit ourselves to the three mentioned proposals (biological, ecological, and phylogenetic species concepts), the availability of different biological ontologies can be easily illustrated. Consider a case of two individuals that can produce fertile offspring but do not inhabit the same ecological niche. Van Valen's classic example for this situation is the North American Oak (1975). *Quercus macrocarpa* and *Quercus bicolor* are commonly considered to be two different species even though they often interbreed and produce fertile hybrid offspring. Is there really a species to which *Quercus macrocarpa* and *Quercus bicolor* belong? Obviously, the biological and the ecological species concepts imply different answers:

(14) There exists a species to which *Quercus macrocarpa* and *Quercus bicolor* belong

(15) There exists no species to which *Quercus macrocarpa* and *Quercus bicolor* belong

A proponent of the biological species concept will accept (14), while a proponent of the ecological species concept will accept (15). It seems that (14) and (15) contradict each other and only one of them can be right. Similar problems arise when we compare the biological and the phylogenetic species concept. Consider the case of the tiger (*Panthera tigris*). The Sumatran Tiger and the Bengal Tiger can produce fertile offspring and according to the biological species concept they belong to the same species. However, Sumatran tigers and Bengal tigers are different enough to form distinct "diagnosable clusters" in the sense of the phylogenetic species concept.[3] Again, we have two contradicting conclusions:

(16) There exists a species to which the Sumatran Tiger and the Bengal Tiger belong

(17) There exists no species to which the Sumatran Tiger and the Bengal Tiger belong

The contradicting conclusions (14) vs. (15) and (16) vs. (17) seem to force us to side with one of the species concepts. This is an unfortunate consequence since we have no idea how to identify the fundamentally correct proposal. At the same time, the discussion of conceptual relativity suggests a solution of the problem. There is

---

[3] For the example *of Panthera tigris*, see also LaPorte (2004, 72). For the distinction between the Sumatran Tiger and the Bengal Tiger based on the phylogenetic species concept, see Jackson (2001). For a critique of the distinction between different tiger species, see Zachos et al. (2013).

no point in discussing whether (14) or (15) and (16) or (17) is fundamentally true, because biologists can work with variety of different but equally acceptable ontologies. The obvious consequence is to claim that (14), (15), (16), and (17) are all true *relative* to different conceptual frameworks:

(14)' There exists$_{biological}$ a species to which *Quercus macrocarpa* and *Quercus bicolor* belong

(15)' There exists$_{ecological}$ no species to which *Quercus macrocarpa* and *Quercus bicolor* belong

(16)' There exists$_{biological}$ a species to which the Sumatran Tiger and the Bengal Tiger belong

(17)' There exists$_{phylogenetic}$ no species to which the Sumatran Tiger and the Bengal Tiger belong

(14)', (15)', (16)', and (17)' do not contradict each other because they only claim that the sentences are true relative to the frameworks of the biological, ecological, or phylogenetic species concepts.

I anticipate two objections to this pluralist strategy. The first one is based on pragmatic considerations, and a biologist might express it along the following lines: "If you're only trying to make the philosophical point that there is not just one fundamental biological ontology, I don't have any objections. But I don't think that your suggestion offers helpful advice to biologists who should work with a consistent ontology across sub-disciplines. Your presentation suggests that biologists who work in different biological fields might find different species concepts helpful. But even if this is true, we should be suspicious of the pluralistic laissez-faire attitude. Biologists in different fields have to be able to talk with each other and it is already easy enough to get lost in translation when talking about taxonomy." I think that a conceptual relativist can accept this objection. Even if there is not only one fundamental biology ontology, one ontology can turn out to be more useful than another. And even if two ontologies turn out to be useful in different contexts, there may be good reasons to push for a consistent ontology across sub-disciplines.

There is another possible objection to the pluralist account of the species debate. Imagine a biologist or philosopher responding in the following way: "When you try to resolve the contradiction between (14), (15), (16), and (17) by differentiating between different ontologies, you not only misunderstand biology, you misunderstand science in itself. Science should not concerned 'different and equally correct conceptual frameworks' but with the fundamental and objective structure of reality. Therefore, the question is not whether a species exists relative to biological, ecological, or phylogenetic conceptual framework. The question is whether that species *really* exists." This appeal to the ideal of exactly one fundamental ontology suggests a reformulation of my examples. Instead of (14)', (15)', (16)', and (17)', we are supposed to discuss the following absolute existence statements:

(14)" There exists$_{absolute}$ a species to which *Quercus macrocarpa* and *Quercus bicolor* belong

(15)" There exists$_{absolute}$ no species to which *Quercus macrocarpa* and *Quercus bicolor* belong

(16)" There exists$_{absolute}$ a species to which the Sumatran Tiger and the Bengal Tiger belong

(17)" There exists$_{absolute}$ no species to which the Sumatran Tiger and the Bengal Tiger belong

(14)", (15)", (16)", and (17)" are, of course, the only "proper ontological" claims in the sense of the project of a fundamental ontology. Philosophers who want to reject conceptual relativity in biology therefore have to insist that biologists should not content themselves with (14)'-(17)' and instead aim for (14)"-(17)" and the question what species *really* exist.

While a proponent of the ideal of one fundamental biological ontology may suggest reformulations in the sense of (14)"-(17)", a closer look at the diversity of biological ontologies renders such a proposal implausible. Let us start with the truism that scientists should believe in the existence of entities that are postulated by their best theories. A pluralist account of the species debate therefore requires that there is not one best theory of species that implies an unambiguous ontology. One way of justifying this claim is to argue that standard criteria of theory choice such as simplicity, empirical adequacy, or explanatory power underdetermine ontological choices in the species debate. However, I think that we can even go a step further. Standard criteria of theory choice do not only underdetermine species ontologies but they imply different ontologies in different explanatory contexts. For example, the biological species concept will have little explanatory value for biologists who are concerned with asexual species while it may be the best choice for many biologists who work, for example, in mammalian biology. What counts as the best account of species will often depend on rather specific features of biological research projects. As Kitcher puts it:

> I have already remarked on the way in which the biological species concept illuminated the issue of the distribution of mosquitoes in the Anopheles maculipennis complex. Yet it should be evident that distinction according to reproductive isolation is not always the important criterion. For the ecologist concerned with the interactions of obligatorily asexual organisms on a coral reef, the important groupings may be those that trace the ways in which ecological requirements can be met in the marine environment and which bring out clearly the patterns of symbiosis and competition. Similarly, paleontologists reconstructing the phylogenies of major classes of organisms will want to attend primarily to considerations of phylogenetic continuity, breaking their lineages into species according to the considerations that seem most pertinent to the organisms under study: reproductive isolation of descendant branches, perhaps, in the case of well-understood vertebrates; ecological or morphological discontinuities, perhaps, in the cases of asexual plants or marine invertebrates. I suggest that when we come to see each of these common biological practices as resulting from a different view about what is important in dividing up the process of evolution we may see all of them as important and legitimate. (1984, 124–25)

Kitcher's comment illustrates the general idea that different accounts of species reflect different explanatory interests. If the choice of an account of species reflects explanatory power and explanatory power varies with explanatory interests, a plurality of explanatory interests will lead to a plurality of species ontologies. In this sense, pluralism is not motivated by a lack of information that underdetermines ontologies but rather by an abundance of information that allows a variety of ontologies with different explanatory strengths and weaknesses. Biological organisms resemble each other in many different ways and different degrees. While some similarities are clearly more interesting than others (no biologist would postulate taxa that are primarily based on color or size), there is always a variety of similarities that can be weighted differently depending on the explanatory interests of scientists. For example, the property of being able to produce fertile offspring is considered crucial by proponents of biological species concept while the property of sharing an ecological niche is considered more important by proponents of the ecological species concept. Given that there is no interest-independent account of the importance of shared properties, there is also no interest-independent account of species or even one general interest-independent biological ontology.

A proponent of the ideal of exactly one fundamental biological ontology may respond by objecting that I have misrepresented the structure of the species debate. While I have suggested that diverse explanatory interests lead to different existence claims, one may object that biologists actually agree on the ontological question what kinds *exist* and only disagree on the question what kinds should be *identified* with species. Instead of a genuine ontological disagreement on what exists, we therefore only have a classificatory disagreement on the best use of the term "species" (cf. Devitt 2011). Instead of the contradicting claims (14)"-(17)", a critic of conceptual relativity may therefore suggest to reconstruct species pluralism along the following lines:

(14)''' There exists$_{absolute}$ a species$_{biological}$ to which *Quercus macrocarpa* and *Quercus bicolor* belong

(15)''' There exists$_{absolute}$ no species$_{ecological}$ to which *Quercus macrocarpa* and *Quercus bicolor* belong

(16)''' There exists$_{absolute}$ a species$_{biological}$ to which the Sumatran Tiger and the Bengal Tiger belong

(17)''' There exists$_{absolute}$ no species$_{phylogenetic}$ to which the Sumatran Tiger and the Bengal Tiger belong

The reformulation (14)'''-(17)''' allows critics of conceptual relativity to accept a diversity of species concepts while still insisting on exactly one fundamental biological ontology. The biological realm provides us with exactly one objective system of natural kinds that "carve nature at its joints" and the only remaining disagreement is what natural kinds deserve the honor of having the label "species" attached to them. This suggestion can be further motivated by the observation that at least some current controversies about biological classification are not concerned with the question what biological kinds exist. For example, consider current controversies about race in biology (Ludwig 2015a). While race theories of the nineteenth and early twentieth century assumed the existence of essentialist and cognitive significant human kinds that turned out to be non-existent, current debates about race in biology are closely entangled with conceptual questions about the use of the term "race" (Ludwig 2014a). Realists in the debate about race identify races with biological entities such as with genetic clusters (e.g. Edwards 2003; Spencer 2014) or clades (e.g. Andreasen 1998), while antirealists (e.g. Glasgow 2008) object that these identifications are based on a misunderstanding of meaning of "race". For example, Hochman (2013) points out that races in non-human biology are usually identified with subspecies while human subpopulations do not meet standard criteria for subspecies. Often, these controversies are not at all concerned with the question what biological kinds *exist* but solely focus on the question whether we should *identify* races with genetic clusters, clades, subspecies, and so on.[4]

While it is certainly correct to point out that some disagreements are only concerned with the best use of labels like "race" or "species", there are also good reasons to make the stronger claim that at least some disagreements are concerned with the existence of biological kinds. For example, many biologists will not only reject the identification of traditional morphological kinds with species but rather reject that morphological species concepts refer to any legitimate biological kinds *at all*. The rejection of morphological accounts of species is therefore not restricted to the use of the term "species" but involves the question whether we should include traditional morphological kinds in biological ontologies.

Similar issues occur in controversies about more recent species concepts as the often highly polarized debates about phenetics and cladistics illustrate. Phenetics developed in the 1960's as an attempt to overcome problems of traditional morphological species concepts by measuring "overall similarity" of organisms on the basis of all available characters. Despite its use of advanced computational methods, phenetics soon came under increasing pressure from cladistic biologists and Quicke's influential *Principles and Techniques of Contemporary Taxonomy* summarizes a common attitude by claiming that phenetic methods "neither provide reliable evidence of evolutionary relationships, nor form a sound basis for classification" (1993, 85). Meier and Willmann (2000, 38) go even a step further by arguing that character-based taxonomies that do not consider reproductive gaps lead to the "the 'creation' of arbitrary species based on arbitrarily chosen sets of characters." Cladistic approaches, however, have come under equally strong criticism by being accused of postulating "taxonomic artifacts" (Zachos et al. 2013) as species. Avines, for example, argues that the cladistic "fanaticism against pheneticism in species concepts" has led to taxonomies that "completely disregard what we do know about underlying ontological reality (in this case regarding such evolutionary processes as mutation, sexual Mendelian inheritance, and genetic recombination)" (2000, 1830).

---

[4] This does not mean that current debates about race are free from more substantial forms of ontological disagreement. In a series of recent papers, Kaplan and Winther have shown that genetic analysis of human subpopulations provides a convincing example of ontological pluralism. First, Kaplan and Winther (2012) develop three technical meanings of genetic variation ("genetic diversity", "genetic differentiation", and "heterozygosity") and show that these different meanings can be combined with a variety of metrics that involve, for example, different sensitivities to allele frequency. Second, Winther and Kaplan (2013, cf. Kaplan and Winther 2014) show how different meanings and metrics correspond to different explanatory interests in a variety of research contexts and therefore lead to the assumption of different biogenomic kinds. In Ludwig (2015a), I distinguish between two types of underdetermination of the ontology of race. The first type is based on quantifier variance and different accounts of what it means for a biological kind to exist. The second type of underdetermination is based on different meanings of "race" and not different meanings of "existence".

The controversies about phenetic and cladistic approaches illustrate that disagreement in the species debate is not limited to the question what taxa deserve the honor of the label "species" but often extends to the question what species concepts refer to legitimate biological kinds *at all*. Furthermore, the observation that the explanatory potential of kinds varies with explanatory interests also allows a more general diagnosis of this situation (cf. 4.4). Similarities between organisms are truly ubiquitous and any scientifically useful account of biological kinds will have to identify relevant biological similarities that constitute legitimate biological kinds. However, any evaluation of the relevance of similarities between organisms will have to acknowledge the diverse explanatory interests of biologists and therefore undermines the idea that we can develop an interest-independent account of natural kinds in biology. The idea of exactly one interest-independent biological ontology that "carves nature at its joints" therefore misunderstands the constitutive importance of scientific interests in the discussion about biological ontologies.[5]

4.2 Cognition

In the last section, I presented the species debate as an example for conceptual relativity in the empirical sciences. Controversies about species concepts do not indicate that biologists have not figured out yet what biological kinds exist but rather that the biological realm can be described in terms of a variety of ontologies that match different explanatory interests. The goal of this section is to argue that the species debate is not a strange exception but illustrates a general phenomenon that is equally common in other scientific disciplines.

If we turn from biology to psychology, we also encounter some rather obvious candidates for conceptual relativity. Maybe the best known examples come from psychiatry and especially from debates about the question what mental disorders exist. Hacking's project of a "historical ontology" (Hacking 2004, cf. Foucault 1966) provides many detailed examples of how our psychiatric ontologies have been shaped by a large variety of epistemic and non-epistemic values. While Hacking's examples clearly challenge the idea of exactly one correct psychiatric ontology (cf. Ludwig 2014c), proponents of the ideal of one fundamental ontology will remain unimpressed. Instead, they will most likely object that conceptual

[5]While the species debate is clearly the best-known ontological controversy in biology, it is not difficult to identify further case studies. For example, Haber and Odenbaugh's special issue on "the edges and boundaries of biological objects" (2009) provides a wide range of interesting examples of ontological debates in biology. Furthermore, recent accounts of systems biology (O'Malley and Dupré 2005) and microbiology (Bapteste and Dupré 2013) also raise ontological issues by arguing that current research is largely concerned with dynamics that may be best described in terms of a process ontology and usually involve entities that belong to different levels of organization in standard object ontologies. Another resource of various examples are debates about the "biological notion of individual" (Wilson and Barker 2013) such as organs (Winther 2011) or entire organisms (Clarke 2010).

relativity in debates about mental disorders disqualifies them as serious ontological debates. Indeed, accounts of mental disorders are shaped by all kinds of social, cultural, and moral considerations and it is by no means surprising that psychiatrists cannot agree on one fundamental psychiatric ontology. However, this only shows that a normative concept such as "mental disorder" has no place in a serious scientific ontology and that we have to limit ourselves to cognitive sciences with more stringent methodological requirements and therefore better prospects to actually carve nature at its joints.

In this section, I will argue that even if we exclude psychiatric ontologies at least for the sake of the argument, conceptual relativity is still ubiquitous in the cognitive sciences and not limited to kinds that have an obvious normative component such as mental disorders. More specifically, I will try to show that the framework of conceptual relativity can be actually applied to the *very concept* of cognition.

One of the currently most passionately discussed issues in philosophy of mind and cognitive science is the question whether cognition extends beyond the organism and is partly constituted by the environment. Proponents of externalism like Clark and Chalmers (1998) argue for an extended account of cognition while internalists insist that human cognition is realized entirely within the brain. The question whether there are extended cognitive processes certainly looks like a substantive issue that leaves little room for compromise or even conceptual relativity. Either human cognition extends beyond the organism or it does not extend beyond the organism – *tertium non datur*. As Rupert (2009, 9) puts it in his discussion of externalism and his own brand of internalism that he calls "the embedded view:""the extended and embedded views are mutually exclusive: it cannot be true both that the human cognitive system consists partly of elements beyond the boundary of the organism (the extended view) and that the human cognitive system is organismically bounded but carries its cognitive work by subtle and complex exploitation of environmental structures (the embedded view)."

My discussion of the role of explanatory interests in scientific ontologies raises doubts about this diagnosis of internalism and externalism as mutually exclusive positions. More specifically, I will argue that different research interests in the cognitive sciences lead to different accounts of cognition. Both externalism and internalism provide useful frameworks in different explanatory contexts and there is no reason to assume that there is only one correct cognitive ontology. The debate about extended cognition provides a further example of conceptual relativity in the empirical sciences. Instead of worrying whether cognition really extends beyond the organism, we should accept that scientists with different explanatory interests may opt to work with different cognitive ontologies (cf. Pöyhönen 2013 for a similar view).

Before I defend this pluralist proposal, it will be helpful to briefly consider the outlines of contemporary debates about externalism. There are several ways of articulating externalism in philosophy of mind and even if we restrict ourselves to Clark and Chalmers' "extended mind hypothesis" (1998), we can still distinguish related but still different claims. Adams and Aizawa (2011, chapter 7), for example, have stressed that there is a difference between the idea of extended cognitive *systems* and extended cognitive *processes*. On the one hand, externalists can claim that the brain and external devices constitute extended cognitive systems. On the other hand, externalists can also assume that neural and external processes compose extended cognitive processes. Arguably, the hypothesis of extended cognitive processes is more ambitious as there could be an extended cognitive system that is composed by the brain and external devices but realizes cognitive processes only in the brain. Furthermore, one can also distinguish between the idea of an extended *mind* and the idea of extended *cognition* by taking the former to be more general than the later. For example, one could assume that the idea of an extended mind also implies some form of extended phenomenal consciousness.

Despite this diversity of formulations, all variants of externalism seem to challenge not only the internalist orthodoxy in philosophy of mind but also our intuitions: isn't it just *obvious* that mind and cognition are in the head not somehow spread out in the environment? What could possibly the reason to endorse such an "outrageous" and "preposterous" (Adams and Aizawa 2011, vii) idea? Most arguments for externalism are based on examples or thought experiments of situations in which external devices play a crucial and indispensable role in cognitive processes. In a second step, externalists argue that the best interpretation of these situations considers external devices as part of cognitive processes. Finally, they conclude that only the "internalist prejudice" (Rowlands 2003, 173) that cognitive processes *must* be exclusively realized in the brain is standing in the way of an externalist interpretation.

Clark and Chalmers' (1998, 12–17) thought experiment of Otto's notebook is probably the best known argument of this kind. Otto suffers from a mild case of Alzheimer's and relies on a notebook as a substitute for his biological memory. When he wants to keep important information, he uses his notebook and is therefore able to access the information at some later point. In Clark and Chalmers' example, Otto wants to go to the Museum of Modern Art (MoMA) in New York. He consults his notebook, retrieves the information that the museum is on 53rd street, and goes to the museum. Clark and Chalmers compare Otto to Inga who also wants to go to the MoMA but does not need a notebook because she has the address stored in her biological memory.

Why are many philosophers inclined to consider the notebook merely a *tool for* Otto's cognitive system while accepting Inga's biological memory as a *part of* her cognitive system? According to Clark and Chalmers, there is no good reason to make this kind of distinction. Of course, there are some obvious differences between Otto and Inga but their information retrieval processes are functionally equivalent in all *systematically important* aspects. For example, both Otto and Inga have access to a reliable information resource that allows them to quickly retrieve the information that the MoMA is located on 53rd street. According to Clark and Chalmers' "parity principle", this kind of functional equivalence ensures cognitive equivalence: "If, as we confront some task, a part of the world functions as a process which, were it done in the head, we would have no hesitation in recognizing as part of the cognitive process, then that part of the world is (so we claim) part of the cognitive process" (1998, 8).

Not everyone is impressed by these thought experiments and proponents of internalism have formulated a variety of objections against the assumption that cognitive processes extend beyond the organism. Internalists often provide alternative interpretations of the externalist thought experiments that are compatible with more traditional positions in philosophy of mind. Rupert (2009, 5), for example, suggests an "embedded view" as an alternative that takes "human cognition to rely heavily on the environment but, nevertheless, to be bounded by the human organism. According to this embedded view, typical cognitive processes depend, in surprising and complex ways, on the organism's use of external resources, while not extending into the environment." Rupert's embedded view accepts some of the crucial lessons from the extended cognition movement without being committed to externalist metaphysics. Adams and Aizawa's (2011, chapter 6) discussion of a "coupling-constitution fallacy" points in a similar direction as it stresses the difference between the claim that human cognition is often coupled with external processes and the claim that human cognition is actually constituted by external processes. Again, this distinction can help to acknowledge the importance of the environment without implying an externalist account of cognition.

Other internalist objections are based on empirical research and especially on the success of internalist assumptions in cognitive psychology (e.g. Rupert 2009, 46; Adams and Aizawa 2011, 58–59). Consider, for example, classical findings in memory research such as Miller's (1956) discovery that the human short-term memory has typically a capacity of plus or minus two units (see also Adams and Aizawa 2011, 9). This finding – and countless other results about memory, perception, attention, and so on – seem to depend on an exclusion of external processes that would "extend" cognition in a way that crucial patterns of neurally realized cognition would not be recognizable anymore.

Given this brief summary of the debate about extended cognition, the prospects of a pluralist account can appear questionable. Indeed, if the debate were only about the correct definition of labels such as "cognition", "cognitive process" or "cognitive system", a pluralist could point out their vague character and urge philosophers to simply accept that cognitive scientists use these labels in a variety of ways. However, both internalists and externalists insist that they are not concerned with definitions but the best theory of cognition. As Adams and Aizawa (2011, 13) formulate it: "We offer this as part of a theory of the cognitive rather than as (part of) a definition of the term 'cognitive.' We do not mean to stipulate that this is just want we mean by 'cognition'." Clark and Chalmers (1998, 10) also argue for the substantive character of the debate: "Thus, in seeing cognition as extended is not merely making a terminological decision; it makes a significant difference to the methodology of scientific investigation. In effect, explanatory methods that might once have been thought appropriate only for the analysis of 'inner' processes are now being adapted for the study of the outer, and there is promise that our understanding of cognition will become richer for it."

A serious defense of pluralist account of the debate about extended cognition therefore has to claim that there are different but equally correct accounts of cognition. In the following, I want to argue for such a pluralism by presenting the debate about extended cognition in analogy to the species debate as a case of conceptual relativity in scientific practice. In the case of the species debate, I argued for a plurality of equally legitimate biological ontologies that reflects diverging explanatory interests of biologists. If we follow standard models of theory choice, considerations of explanatory potential should be of crucial concern in ontological choices. However, the explanatory potential of species concepts varies with the explanatory interests of biologists and a diversity of explanatory interests in biology therefore leads to a diversity of biological ontologies. In the following, I want to argue for an analogous situation in the debate about extended cognition. Cognitive scientists often have different explanatory interests that come with different conceptual needs. In the context of some research projects, it is helpful to restrict the cognitive realm to neurally realized structures and to deny the existence of extended cognitive processes. In the context of other research projects, the assumption of extended cognitive processes is very helpful and there is no reason to restrict ourselves to internally realized entities.

In order to make the case for internalism in cognitive science, let us consider the example of Miller's (1956) landmark paper on the capacity of the short-term memory. If we would subscribe to an externalist account of the short-term memory and allow different kinds of memory extensions, we would distort Miller's famous result that the human short-term memory has a distinctive capacity of seven plus minus two units. As Adams and Aizawa (2011, 63–68) point out, examples of this kind are legion. Large parts of contemporary memory research are concerned with patterns that are unique to processes that are realized in the human brain. Consider, for example, standard effects in cognitive psychology such as priming, masking, primacy and recency effects. Many experiments on human short-term memory ask subjects to recall a list of words or other items immediately after learning them. These experiments reveal distinct patterns of the human short-term memory. For example, subjects tend to remember the first (primacy effect) and the last (recency effect) items in a list better than the others. Also, it is usually possible to increase performance with respect to specific items through priming or to decrease the performance through masking (see Baddeley 2007 for an abundance of examples). These kinds of effects are not only of crucial importance for psychologists who investigate how human memory works but they are also bound to the unique structure of the human brain. An artificial "extension" of human memory would not exhibit the same patterns unless it was specifically designed to mimic human biological memory.

Given these examples, we can formulate a simple argument in favor of an internalist account of cognition: (1) Large parts of memory research are concerned with patterns that are specific to neurally realized processing. (2) An extended account of cognition would distort these patterns by treating externally stored information sources with fundamentally different retrieval patterns as memory. (3) Memory researchers should not adopt an account of cognition that distorts their research. ∴ Memory researchers should not accept an extended account of cognition.

How should externalists react to this argument? One option is to reject the second premise according to which an extended account of cognition would distort the patterns memory researchers are concerned with. Although externalists argue that human memory can extend beyond the boundaries of the organism, they are by no means committed to the claim that there is nothing unique about neural processing and neurally realized cognitive processes. On the contrary, it is obvious that biological and non-biological processes are different in many ways and we have every reason to expect these differences to have a measurable impact. Externalists can argue that they do not want to deny any of this and that they are also happy to accept that many psychologists are interested in unique aspects of neurally realized memory. Externalism only seems to require a different interpretation of this situation: psychologists who are concerned with unique aspects of neurally realized cognition are not concerned with human cognition *in general* but only the *biological part* of human cognition.

Externalism may be consistent with this kind of memory research but consistency is not enough to convince psychologists. On the contrary, it seems obvious that psychologists who work on unique aspects of neurally realized processing will consider it completely implausible that they are not concerned with human memory and cognition in general but *only with the biological parts* of human memory and cognition. This is not only an intuitive issue as these psychologists could rightly point out that the externalist interpretation is not only useless but actually an unnecessary complication that adds an additional step without providing any benefits for their research.

Ironically, one can formulate this argument in direct analogy to one of Clark and Chalmers' arguments *in favor* of externalism. In their thought experiment of Otto's notebook, Clark and Chalmers discuss the internalist proposal that Otto does not actually know the address of the MoMA but only knows that the address is in his notebook: "The alternative is to explain Otto's action in terms of his occurrent desire to go to the museum, his standing belief that the Museum is on the location written in the notebook, and the accessible fact that the notebook says the Museum is on 53rd Street; but this complicates the explanation unnecessarily. If we must resort to explaining Otto's action this way, then we must also do so for the countless other actions in which his notebook is involved; in each of the explanations, there will be an extra term involving the notebook. We submit that to explain things this way is to take *one step too many*" (Clark and Chalmers 1998, 13).

Clark and Chalmers' argument is based on an appeal to the value of simplicity of theories: we should not make theories unnecessarily complicated and only choose a more complicated theory if there is some substantial explanatory payoff. It seems that many psychologists who are mostly concerned with unique aspects of neurally realized processing could turn this argument against Clark and Chalmers. The idea of extended cognition makes the situation unnecessarily complicated and asks them to take *one step too many*.

Even if some areas in cognitive psychology suggest an internalist framework, externalists can still object that the case for internalism suffers from a focus on one-sided examples as there is also a lot of empirical evidence that favors an externalist analysis. And indeed, many externalists are convinced that contemporary research on problem solving provides strong arguments for an externalist framework.

Gray and Fu's (2004) study, for example, is based on the assumption that "knowledge can be in-the-world or in-the-head" (359) and tries to answer the question under what circumstances subjects utilize different sources of knowledge in problem solving strategies. In a series of experiments, Gray and Fu asked subjects to use a virtual control panel of a VCR player to record a television show. The information that was needed to program the VCR (show name, start/end time, day, channel, and program number) was available on the computer screen. Furthermore, they tested three different scenarios: In the first scenario, the information that was needed to program the VCR was clearly visible on the screen and the subject had free access to the information (*Free-Access* scenario). In the second scenario, the information was covered by a gray box and the user had click on the target with a mouse to bring the information to the foreground (*Grey-Box* scenario). The third scenario allowed subjects to memorize the necessary information in advance and further obscures the window with the VCR player as soon as the window with the programming information is accessed (*Memory-Test* scenario).

Gray and Fu found a diversity of problem solving strategies in the different scenarios that were dependent on the time that was needed to access the necessary information. Subjects prefer the most efficient strategies that minimize time investment no matter whether the strategy relies on in-the-world or knowledge in-the-head. Gray and Fu conclude that there is no "privileged status of any location or type of operation" (Gray and Fu 2004, 378, 380) and describe this tendency to minimize time investment independently of the location of the knowledge as a "soft constraint" in interactive behavior.

This study provides an excellent example of externalism-friendly research in cognitive science that suggests that knowledge can be realized both in the brain and in the external world. An extension of the short-term memory to external devices may distort the patterns of Miller's (1956) famous study but a restriction of knowledge to biologically realized knowledge would distort the patterns of Grey and Fu's study in the same way. If we do not even allow knowledge in-the-world, then we cannot detect the interesting soft constraints that determine which knowledge source and problem solving strategy is most likely to be selected.

It is not difficult to predict the internalist response to these experiments. Recall my brief presentation of Rupert's embedded account of cognition as well as Adams and Aizawa's discussion of the "coupling-constitution fallacy". According to these internalist accounts, human cognition is often coupled with and dependent on external processes but cognitive processes are not literally constituted by these external processes. Internalists will stress that the experiments are actually consistent with these internalist alternatives: they can accept that Gray and Fu's soft constraints illustrate a surprising analogy between information access in-the-world and knowl-

Table 4.1 Summary of the dialectical situation

The case for internalism	The case for externalism
Argument: Experiments such as Miller (1956) presuppose an internalist framework. An extension of the short-term memory beyond the brain would distort crucial patterns.	*Argument*: Experiments such as Gray and Fu (2004) presuppose an externalist framework. A restriction to brainbound cognitive processes would distort crucial patterns.
Objection: Experiments such as Miller (1956) are consistent with externalism but require a distinction between a biological short-term memory (that has a capacity of 7 ± 2 units) and general short-term memory.	Objection: Experiments such as Gray and Fu (2004) are consistent with internalism but require a distinction between information in-the-world that is used in the same way as knowledge and genuine knowledge that is realized within the brain.
Rebuttal: Consistency is not enough and the externalist proposal asks cognitive psychologists to make an unnecessary distinction. This is one step too many.	Rebuttal: Consistency is not enough and the internalist proposal asks cognitive psychologists to make an unnecessary distinction. This is one step too many.

edge retrieval in-the-head. In many problem solving scenarios, it does not matter whether information is stored in the brain or in the external world and subjects simply choose the most efficient strategy. Internalists do not have to reject these findings but only insist on a different metacomment: Information in-the-world is not literally knowledge in-the-world, but only *becomes* knowledge after it has been accessed and stored in the biological memory.

While internalists are certainly correct in pointing out the availability of non-externalist interpretations, this move creates the same problems as externalist interpretations of psychological research that is concerned with unique aspects of neurally realized cognition: these interpretations may be consistent but they add unnecessary complications to the discussion of the experiments. Gray and Fu could object that the internalist interpretation requires a distinction between the real (aka biologically stored) knowledge and mere information that becomes knowledge at some later point without generating any explanatory benefits for their research projects. The internalist interpretation requires an additional step that is of no use for Gray and Fu and that is, again, *one step too many*.

To sum up: internalists and externalists can both present research in cognitive psychology that favors their conceptual proposals. There are studies such as Miller (1956) that favor an internalist account and studies such as Gray and Fu (2004) that favor an externalist account. As Table 4.1 illustrates, internalism and externalism end up in almost identical dialectical situations:

So far, I have tried to show that some areas in psychology suggest an internalist framework while others suggest an externalist framework. Although I do not want to argue that cognitive psychology neatly separates into an internalist and an externalist part, I do think that there is a more general pattern. Recall some examples that seem to favor internalism such as research on the capacity of the short term memory or on effects such as priming and masking. These examples belong to a core program of cognitive psychology as it emerged from the "cognitive revolution" in the middle of the twentieth century (e.g. Baars 1986). The cognitive revolution is often characterized as an attempt to (re-)enter the black box and to move beyond the behaviorist analysis of input–output relations. Early cognitive psychologists argued that successful psychology has to consider the internal machinery of cognition and that the uncovering of these mechanisms should be a core task of their discipline. I think that there is a clear sense in which this program suggests an internalist framework: the venture inside the "black box" of the brain leads to an uncovering of mechanisms that typically reflect their unique neural realization. Examples include but are by no means limited to the unique capacity limits of memory and perception, unique breakdown patterns in neurology, or unique effects such as priming and masking. External mechanisms typically exhibit very different patterns which make it attractive for many psychologists to restrict cognitive processes to neurally realized processes.

Although this kind of research on the internal machinery of cognition has been a defining aspect of the cognitive revolution, cognitive psychology has always included a larger variety of projects. Consider, for example, research on problem solving strategies that is not so much concerned with properties of internal mechanisms but with human behavior and strategies in complex environments. Problem solving usually relies on internal and external resources and humans often utilize both kinds of resources in surprisingly similar ways. Gray and Fu's experiments provide great examples as they show how little the external-internal distinction matters in the context of certain problem-solving tasks. This kind of problem solving research often suggests a very different approach to cognition as it is concerned with processes that rely on the active role of the environment without having any use for the internal-external distinction.[6]

My discussion suggests the following picture. Some areas in cognitive psychology are mostly concerned with unique aspects of neurally realized cognition and have little or no use for the idea of extended cognition. Other areas that are concerned with human action in complex environments have little or no use for the internal-external distinction. The debate about extended cognition therefore provides another example for conceptual relativity in the empirical sciences. Cognitive scientists with different explanatory interests will find different patterns meaningful and will therefore work with different cognitive ontologies. Instead of wondering whether extended cognitive processes really exist, I suggest that we should accept that different research projects in cognitive science will use different cognitive ontologies.

---

[6]The situation becomes more complicated when we consider other areas of research that are also concerned with cognition in one way or another. For example, I have argued (Ludwig 2013, cf. Ludwig 2014b) that science communication studies should rely on externalist accounts of cognition and knowledge.

4.3   Intelligence

So far, my examples of conceptual relativity have been concerned with very general categories in biology and cognitive science. In my third and last case study, I want to show that conceptual relativity can also occur in more specialized debates by focusing on the question whether one specific entity – general intelligence – exists. Debates about the existence of a general intelligence have a prominent place in psychology since the early twentieth century. In 1904, the British psychologist Charles Spearman published an article in which he introduced factor analysis as a method to "objectively measure" intelligence (Spearman 1904, cf. Horn and McArdle 2007). According to Spearman, the positive correlations between different cognitive abilities indicate the existence of a general intelligence factor. Although Spearman's assumption of a general intelligence factor became highly influential in psychological research, it was soon criticized by psychologists like Louis Leon Thurstone as a misinterpretation of psychometric data. Thurstone argued that a more satisfying interpretation of intelligence tests would not assume one general intelligence factor but several primary mental abilities such as verbal comprehension or numerical ability (Thurstone 1938).

Ever since the debates about Thurstone's *primary mental abilities*, psychologists passionately disagree on the issue of general intelligence. One the one hand, there is a camp of psychologists who often work in the psychometric tradition and consider the idea of a general intelligence crucial for their research on issues such as intelligence differences in human populations or the impact of environmental factors on intelligence. On the other hand, many psychologists have questioned the existence of one general intelligence and instead favor a plurality of intelligences. And indeed, it is not difficult to cast doubt on the viability of a general intelligence concept: people clearly have different cognitive abilities and are highly intelligent with respect to some tasks and less intelligent with respect to others. Some people are good with visual orientation but bad with language. Others are great musicians but lousy mathematicians. Does it not, therefore, make more sense to assume different abilities or different intelligences than the existence of one general intelligence?

Neurological case studies provide another reason to doubt the usefulness of the assumption of a general intelligence in psychology. Many cognitive abilities can be dissociated from each other through brain damage and these cases of double dissociation seem to indicate that there is no such thing as a general cognitive ability that could be called "intelligence." Instead, there are many cognitive abilities that are largely independent from one another. Another evidence for the assumption that cognitive abilities are often independent from each other are exceptional cases, such as the savant syndrome, in which a person has an area of unmatched expertise that starkly contrasts with the individual's overall limitations. Again, a general intelligence concept appears ill-suited to deal with these observations as it is not clear how we should think of a general intelligence given the often modular organization of human cognition. These observations seem to indicate that the assumption of a

Table 4.2   Tasks of the WPCT

Test task	
1. Reading	Answer questions about a paragraph
2. Vocabulary	Choose synonyms for a word
3. Grammar Identify	correct and poor usage
4. Quantitative Skills	Read word problems and decide whether they can be solved
5. Mechanical reasoning	Examine a diagram and answer questions about it; requires knowledge about mechanical and physical principles
6. Spatial Reasoning	Indicate how two-dimensional figures will appear if they are folded through a third dimension
7. Mathematical achievement	A test of high school algebra

From Anderson (2000, 445)

general intelligence presupposes the flawed idea of a centralized human cognition that has been challenged by the empirical literature.

Proponents of a general intelligence concept have a standard reply to these objections and point out that cognitive abilities are not independent from each other, as intelligence tests illustrate. Cognitive abilities are positively correlated with one another; people who perform above average with respect to one task also tend to perform above average with respect to other cognitive tasks. Correlations between cognitive abilities show that a general intelligence concept does not lump completely different things together. Rather, it reflects that our different cognitive abilities rest on shared cognitive resources. "General intelligence" simply refers to these shared cognitive resources (Jensen 1998).

Consider the Washington Pre-College Test (WPCT) as a simplified illustration of how the different cognitive abilities are correlated.[7] The WPCT tests six different abilities (Table 4.2).

A well-established result is that the performances between the different tasks are positively correlated. Table 4.3 shows the intercorrelations between the results of the tests in Table 4.2.

According to proponents of a general intelligence concept, the intercorrelations show that different cognitive abilities rest on a shared cognitive resource, which is the general intelligence. Is this a convincing argument? Many psychologists do not think so, and the example of WPCT is also helpful in illustrating their doubts. Consider the following representation of the results.

While performances in some tasks such as 1 and 2 are strongly correlated, correlations between other performances such as 1 and 6 are rather weak. Figure 4.1 offers a two-dimensional visualization of this, in which the distance between the dots represents the strength of the correlation. It becomes obvious that there are clusters, as the linguistic tasks and the mathematical tasks are very close together. Should we not interpret these results as evidence for different intelligences such as a linguistic and a mathematical intelligence?

---

[7] The presentation of the WPCT follows Hunt (1985) and Anderson (2000, 444–448).

Table 4.3 Correlation coefficients of different tasks in the WPCT

Test Nr.	1	2	3	4	5	6	7
1	1.00	.67	.63	.40	.33	.14	.34
2		1.00	.59	.29	.46	.19	.31
3			1.00	.41	.34	.20	.46
4				1.00	.39	.46	.62
5					1.00	.47	.39
6						1.00	.46
7							1.00

From Anderson (2000, 445)

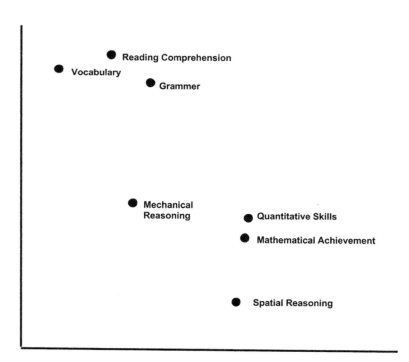

Fig. 4.1 Test results in a two-dimensional space. Distance between points represents intercorrelations (Adapted from Anderson (2000, 446))

Psychologists do not agree on how to answer this question, with proponents of a general intelligence concept and advocates of a multiple intelligence approach apparently contradicting each other with the following claims:

(18) There is just one general intelligence
(19) There are several intelligences

Who is right? It is generally acknowledged that the statistical methods of psychometrics, such as factor analysis, cannot solve the issue. While the general intelligence concept is traditionally tied to factor analysis and the construction of a general intelligence factor, it is equally possible to postulate several intelligence factors and to use factor analysis in support of (19).[8] From a statistical point of view, (18) is as good as (19). But if factor analysis does not solve the issue, how can we decide whether there are one or several intelligences? In his classical book, *Frames of Mind*, Howard Gardner defends a theory of multiple intelligences and presents six criteria to evaluate if something qualifies as an intelligence:

1. the potential for brain isolation by brain damage;
2. its place in evolutionary history;
3. the presence of core operations;
4. susceptibility to encoding (symbolic expression);
5. a distinct developmental progression;
6. the existence of idiot-savants, prodigies and other exceptional people;
7. and support from experimental psychology. (Gardner 1985, 59–69).

The application of these criteria led Gardner to the assumption of six different intelligences: linguistic, musical, mathematical, spatial, bodily kinesthetic, and personal. However, proponents of the general intelligence concept will not be convinced by Gardner's proposal. First of all, they do not need to accept Gardner's criteria, and many proponents of a general intelligence concept criticize his proposal of a crude package of limited systematic value (Waterhouse 2006, 207–213). Second, proponents of a general intelligence concept will deny that musical or personal abilities have anything to do with intelligence.

Given the earlier discussion of conceptual relativity, it will not come as surprise if I suggest that there is not only one correct answer to the question how many intelligences exist. Empirical evidence will not settle the issue as proponents and critics of the general intelligence concept can agree on all the data from psychometric research and neuroscience and still disagree on (18) and (19). If an appeal to empirical adequacy does not settle the issue, the debate will have to consider "super-empirical" factors which bring us back to the interest-relativity of criteria such as explanatory power. Cognitive scientists who work at the intersection of cognitive psychology and cognitive neuroscience, for example, often prefer to work with more specific cognitive entities than "general intelligence" that are more that are likely to be correlated with specific pathways of neural processing. While work at the intersection of cognitive psychology and cognitive neuroscience tends to lead to more fine-grained accounts of cognitive processes, other areas of psychology have more use for a general intelligence concept. This is especially obvious in the case of traditional psychometric research that is often concerned with general differences between populations and presupposes a general intelligence concept. Again, a debate about the question how many intelligences *really* exist misses that researchers with different explanatory interests will find different patterns meaningful.[9]

---

[8] Factor analysis was introduced by Spearman (1904). See Gould (1996, 269–285) for an opinionated introduction to the history of intelligence research and the use of factor analysis.

[9] My discussion of intelligence is largely analogous to the discussion of species and extended cognition. Different explanatory interests within a scientific discipline lead to different conceptual

4.4 What about Natural Kinds?

So far I have argued that conceptual relativity is common in the empirical sciences. In disciplines such as biology or the cognitive sciences, we often encounter a plurality of equally legitimate ontologies that reflect a plurality of explanatory interests in science. There is not only one correct answer to the question what species exist, not only one correct answer to the question whether extended cognitive processes exist, and not only one correct answer to the question how many intelligences exist.

Many philosophers will respond to this presentation by objecting that conceptual relativity has unacceptable philosophical consequences and leads to a radical philosophical anti-realism or relativism. One way of specifying this objection is to argue that conceptual relativity undermines the status of scientific kinds as natural kinds. However, there is hardly a canonical notion of "natural kind" in contemporary philosophy. As Hacking (2007, 203) puts it: "There no longer exists what Bertrand Russell called 'the doctrine of natural kinds'—one doctrine. Instead we have a slew of distinct analyses directed at unrelated projects." In the evaluating the objection that conceptual relativity leaves no room for natural kinds, we will therefore need a more fine-grained account of what it means for a kind to be "natural".

In the following, I will argue that conceptual relativity is indeed incompatible with the ideal of natural kinds that are completely independent of contingent explanatory interests. A rejection of completely interest-independent natural kinds, however, does not imply that scientific kinds reduce to conventions or to pragmatically useful kinds. Instead, philosophers of science have suggested a large range of more moderate accounts of natural kinds (e.g. Boyd 1999; Ereshefsky and Reyodon 2015; Franklin-Hall 2015; Häggqvist 2005; Khalidi 2013; Ludwig 2015b; Magnus 2012; Slater 2014; Wilson et al. 2007). While I will not discuss every single proposal, I will argue that conceptual relativity is compatible with two of the core observations that motivate moderate accounts of natural kinds: (1) scientific kinds

---

needs and it is by no means surprising that some psychologists prefer to work with an ontology that only includes a general intelligence while others insist on an ontology of multiple intelligences. However, the case of intelligence adds a further complication to the debate about scientific ontologies as different accounts of intelligence do not only reflect different explanatory interests but also often non-epistemic (e.g. moral, educational, political) values. Gardner's theory of multiple intelligences, for example, has become tremendously popular with pedagogues who feel that the concept of a general intelligence has proven harmful in education. Instead of classifying people with a general intelligence scale, the theory of multiple intelligences allows educators to concentrate on individual strengths and weaknesses (e.g. Chen 2004, 20–22). Furthermore, the question whether psychologists should continue to use a general intelligence concept is also often answered differently based on different attitudes towards research on cognitive differences between human populations. While proponents of this kind of research occasionally point out its alleged use in public policy contexts (e.g. Herrnstein and Murray 1994, Part IV), critics consider it a dangerous platform for pseudo-scientific justifications of racism and sexism (e.g. Gould 1996). At least in the case of the intelligence debate, the choice of a specific intelligence concept seems to be closely entangled with social values and raises further question about the role of values in scientific practice. While I assume that normative considerations often play an important role in ontological choices of the human sciences (e.g. Ludwig 2014a, 2015b), this claim is not necessary for a general defense of conceptual pluralism.

often reflect the empirical discovery of property clustering and (2) they support diverse inductive inferences.

Natural Kinds as Interest-Independent Kinds Conceptual relativity in a domain such as biology is clearly incompatible with a strong natural kind realism that insists that biological kinds are completely independent of contingent explanatory interests. In the case of the species debate, I argued that scientists with different explanatory interests will find different biological patterns meaningful and will therefore postulate different biological kinds. A proponent of the ideal of interest-independent biological kinds would therefore have to provide a strategy of stepping behind the plurality of interest-dependent taxonomies in scientific practice and propose an alternative strategy of evaluating what biological kinds objectively exist.

This would be a realistic strategy if members of a species were distinguished by intrinsic essences. Let us assume for the sake of the argument that members of the same species share a common intrinsic structure (e.g. the genome) that clearly distinguishes them from members of other species. In this case, it would seem plausible to postulate a privileged taxonomy that "carves up" the biological realm along its discrete and natural discontinuities. However, the obvious problem with this assumption is that the biological realm does not have such a discrete structure and that the plurality of species concepts is a direct result of the plurality of epistemically fruitful ways of "carving up" the biological realm.

If there are no intrinsic essences that clearly separate biological taxa, we have to engage with the diverse biological (e.g. genetic, phylogenetic, morphological, physiological, ecological, behavioral...) properties that distinguish different species. This puts a strong natural kind realism in an uncomfortable position as it is not easy to see how the relevance of shared properties for kind membership could be evaluated independently from our explanatory interests. As far as I can see, a proponent of interest-independent biological kinds has two options. On the one hand, the evaluation could be purely quantitative. The more properties individuals have in common, the more natural is the kind to which they belong. On the other hand, an evaluation could be qualitative, where the relevance of shared properties would depend on some kind of assessment of their importance.

It seems evident that no purely quantitative model will be successful. The most obvious problem is that objects always share countless artificial and gerrymandered properties (e.g. Lewis 1983). Consider electrons and Tasmanian devils. We can easily construct an artificial kind "electron-or-Tasmanian-devil" and point out that every member of this kind has the property of:

- either being an electron or a Tasmanian devil
- either being an electron or a living being with four legs
- either being a subatomic entity or being native to Tasmania
- not being a Tasmanian Tiger
- not being the King of France.
- being smaller than the Eiffel tower
- being smaller than two Eiffel towers
- ...

If we allow gerrymandered properties, there cannot be a quantitative measurement of the relevance of shared properties because two objects will always have infinite properties in common. Even if we would find a way to exclude all gerrymandered properties and to quantify over legitimate properties, however, it would remain highly doubtful that we can quantitatively measure the naturalness of a kind by counting the number of shared properties.

One problematic consequence of a purely quantitative account would be that more specific kinds would always be more natural than general kinds. For example, compare the kind "mammal" with the kind "mammal with four legs." "Mammal with four legs" is a slightly more specific kind as its members have all properties of mammals and the additional property of having four legs. Or compare "tiger" with "tiger in the San Diego Zoo." Tigers in the San Diego Zoo certainly share a number of unique properties – not only shared geographic location but arguably also similar nutrition, similar behavioral patterns that are the effect of captivity, and so on. If we would measure the relevance of shared properties on a purely quantitative basis, we would arrive at the odd conclusion that kinds like "mammal with four legs" and "tiger in the San Diego Zoo" are more natural than kinds like "mammal" or "tiger" because their members share more properties. Clearly this is unacceptable for any theory of natural kinds that takes scientific practice seriously and illustrates that we cannot understand naturalness by quantifying over the properties that members of a kind share.

The examples of mammals and tigers illustrate the more general point that a purely quantitative account of naturalness would lead to "natural kinds" that are completely unacceptable for scientists. Or, to put it the other way around: any useful account of scientific kinds must be based on an evaluation of properties as more or less important. Consider, for example, the fact that males of many bird species appear more similar to males of other bird species than to females of the same bird species (Ereshefsky 2001, 104). Or consider that early-stage embryos of mammals appear more similar to each other than to adults of their species. Or consider the status of monophyletic higher taxa as scientific kinds (cf. Rieppel 2005; Boyd 2010) despite staggering differences between their members. In order to justify their ontologies, biologists often have to move beyond counting properties and instead identify important properties such as a shared phylogeny.

If there is no purely quantitative measure of naturalness, any comprehensive account of natural kinds needs to evaluate the importance of shared properties. Of course, this is what scientists do all the time in the identification of scientific kinds. Proponents of different species concepts typically do not disagree on what properties individuals have, but rather on how relevant they are. Recall the example of *Quercus macrocarpa* and *Quercus bicolor*, who can produce fertile offspring but inhabit different niches. Proponents of the biological and ecological species concept will agree on the properties of the individuals but disagree over whether being able to interbreed or inhabiting the same niche is *more relevant*.

To sum up, conceptual relativity does indeed contradict a strong metaphysical interpretation of natural kinds. The claim that conceptual relativity extends to the empirical sciences implies that there are different but equally fundamental ways of identifying natural kinds. What counts as a natural kind also depends on the explanatory interests of scientists. We can therefore formulate the basic argument along the following lines:

1. Whether or not a kind is a natural kind depends on whether the members of a kind share sufficiently relevant properties.
2. There is no quantitative way to measure the relevance of shared properties.
3. If there is no quantitative way to measure the relevance of shared properties, then their relevance also depends on the explanatory interests of scientists.
∴ Whether a kind is a natural kind also depends on the explanatory interests of scientists.

It is important to interpret this conclusion carefully. My claim is not that the naturalness of a kind depends *only* on explanatory interests of scientists but that it *also* depends on the interests of scientists. The conclusion therefore contradicts a strong natural kind realism but does not imply an equally strong anti-realism. Contrary to a strong natural kind realism, I have argued that we cannot think of natural kinds as completely non-epistemic. However, this does not mean that scientific kinds reduce to interesting or pragmatically useful kinds. On the contrary, I will argue that a look at more moderate notions of natural kinds suggests that conceptual relativism is compatible with a broadly realist outlook that understands scientific kinds not only in terms of pragmatic interests but also in terms of the empirically discovered structure of the biological realm.[10]

Moderate Notions of Natural Kinds I have argued that conceptual relativity in scientific practice undermines accounts of natural kinds as interest-independent kinds. One possible reaction is to drop the notion of natural kind altogether (Hacking 2007, cf. MacLeod and Reydon 2013). Kitcher (2008, 119), for example, argues that there is "no higher standard to which our concepts are to answer than the efficient satisfaction of the purposes of inquiry; those purposes are set, not by nature, but by us". However, it is also possible to use considerations of explanatory interests as a starting point for the development of moderate notions of natural kinds that do not rely on interest-independence (cf. Ludwig 2015b).

Even if we primarily think of natural kinds in terms of epistemic needs of scientists (cf. Love 2009; Brigandt 2011), it seems reasonable to ask *why* some kinds match the explanatory interests of scientists better than others. For example,

---

[10] The realism issue can easily lead to confusion due to the many possible interpretations of "realism". For example, many philosophers who stress the interest-dependency of scientific kinds (e.g. Stanford 1995; Ruphy 2010; Franklin-Hall 2015) endorse the label "anti-realism" even if they do not want to claim that scientific kinds reduce to conventions. For example, Franklin-Hall (2015) argues that "it is customary—and I believe well-motivated—to understand realist views as those maintaining the full objectivity or mind-independence of the natural kinds." Of course, this is merely a terminology point but a definition of natural kind realism in terms of "full objectivity" and anti-realism in terms of "everything else" seems prone to create misunderstandings. In contrast, I will assume that there is a gradual spectrum in between the realist extreme of full objectivity and the anti-realist extreme of pure conventionalism. In this sense, I take conceptual relativity to be compatible with a moderate realism regarding natural kinds.

consider the contrast between a species concept that is actually used in contemporary biology (e.g. Mayr's biological species concept) and another artificial (e.g. purely color-based) species concept. It is not difficult to explain why the former is epistemically more useful than the latter. The ability to interbreed is of crucial importance in ensuring gene flow between populations while reproductive isolation leads to increasing differences between populations over time. Mayr's biological species concept defines species in terms of interbreeding therefore relies on a mechanism that is of central relevance for understanding various similarities and differences between populations. In contrast, being of the same color is not a comparable mechanism and does not indicate meaningful biological similarities.

There are a number of related lessons to learn from the contrast between epistemically useful scientific kinds and artificial kinds such as "color-based species." First, members of epistemically useful kinds typically share a large number of relevant properties. For example, members of a species share a wide range of anatomical, behavioral, genetic, phylogenetic, physiological properties while members of an artificial color-based species (say "brown animals") only share the relevant property of being brown. Second, these properties are not randomly distributed but clustered in a way that they support diverse inductive inferences (e.g. Magnus 2012; Häggqvist 2005). For example, if we learn that an animal x is a tiger, we can make a lot of inferences about x. In contrast: if we learn that x is a member of the artifical taxon "brown animal," we can only infer that x is brown. Furthermore, the stable clustering of properties also implies that knowledge about some subclusters allows inductive inference to other clusters (cf. Slater 2014). For example, if we know that x has the typical anatomy of a tiger, we can infer that x will probably also exhibit typical tiger behavior.

The observation that scientific kinds often involve property clusters that support inductive inferences is ubiquitous in current debates about natural kinds. The orthodox interpretation of this observation in the life sciences is based Boyd's homeostatic property cluster (HPC) theory (cf. Boyd 1999; Rieppel 2005; Wilson et al. 2007). HPC-theories do not only point out stable property clustering but also explain clustering of properties as "the result of what may be metaphorically (sometimes literally) described as a sort of homeostasis. Either the presence of some of the properties in F tends (under appropriate conditions) to favor the presence of others or there are underlying mechanisms or processes which tend to maintain the presence of the properties in F, or both" (Boyd 1999, 143).

Boyd's HPCs provide an attractive framework for the discussion of biological taxa. Indeed, there are underlying mechanisms that cause members of the same species to share many systematically important properties. In the case of sexual species, interbreeding is an obvious mechanism that explains property clustering among the members of a species. A comprehensive story of the causes of property clustering in species would have to be more complex, as there are other sources of evolutionary stability that apply to nonsexual as well as sexual species (Millikan 1999, 54). But even without a detailed biological story, HPC seems to provide a helpful approach to natural kinds that is applicable to many biological examples.

Boyd (1999, 82–84) suggests to extend this cluster kind realism from biological taxa to kinds in other disciplines such as mineral types, galaxy types, geological formations, storm patterns in meteorology, and so on. All of these entities seem to be constituted by property clusters that are based on shared underlying mechanisms. One may also be inclined to extend this account to my case studies of extended cognition and intelligence. Concepts of cognitive processes and intelligence(s) may also refer to property clusters with shared underlying mechanisms and therefore turn out to be HPCs.

While the idea of homeostatic mechanisms is often illuminating in discussions of scientific kinds, there are good reasons to doubt that we can *define* natural kinds in terms of HPCs. On the one hand, HPCs do not seem to be necessary for natural kinds as there are many important scientific kinds that do not match Boyd's characterization. For example, it has often been noted that entities in particle physics do not constitute homeostatic property clusters (e.g. Magnus 2014). Furthermore, Ereshefsky and Reydon (2015) provide a long list of (e.g. non-causal and functional) kinds that do not fit Boyd's characterization. On the other hand, one may also wonder whether HPCs are sufficient for natural kinds. For example, data driven research in the life sciences allows the discovery of countless rather uninteresting property clusters and it is at least not immediately clear that any cluster that satisfies HPC must be considered a natural kind.

However, the goal of this section is not endorse one general account of natural kinds but to argue that the observations that motivate moderate accounts of natural kinds (e.g. property clustering, homeostatic mechanisms, projectability, epistemic relevance) are compatible with conceptual relativity. For example, property clustering, homeostatic mechanisms, projectability, and epistemic relevance are arguably satisfied by a variety of species concepts and therefore compatible with thorough pluralism regarding species ontologies.

Furthermore, one may argue that criteria such as "property clustering" or "homeostatic mechanisms" are themselves relative to explanatory interests of scientists. For example, Craver (2009, 575) has argued that "conventional elements are involved partly but ineliminably in deciding which mechanisms define kinds, for deciding when two mechanisms are mechanisms of the same type, and for deciding where one particular mechanism ends and another begins. This intrusion of conventional perspective into the idea of a mechanism raises doubts as to whether the HPC view is sufficiently free of conventional elements to serve as an objective arbiter in scientific disputes about what the kinds of the special sciences should be." Even more generally, one can argue that there is not an interest-independent answer to the question at what point shared properties constitute a cluster in the sense of moderate accounts of natural kinds. For example, it seems reasonable to require that property clustering is stable in the sense that the presence of some properties allows inferences to the presence of others. However, it is not clear that we can provide an account of stableness that is independent of the unique requirements of disciples that use the kinds in question (cf. Slater 2014).

To sum up, my discussion of natural kinds has involved a negative and a positive claim. On the one hand, I have rejected a strong metaphysical notion of natural kinds that is built on the ideal of complete interest-independence. On the other hand, I have argued that the rejection of interest-independence is compatible with observations that motivate more moderate accounts of natural kinds such as the empirical discovery of property clustering, the fact that scientific kinds typically support diverse inductive inferences, and the existence of homeostatic mechanisms.

Of course, there is still a lot of room for disagreement regarding natural kinds but I do not think that my arguments for conceptual relativity need to be tied to one specific notion of natural kinds. On the one hand, I am sympathetic with Dupré's (2002) claim that "natural kind is not a natural kind term" and that we will probably not find one unifying notion of natural kinds that is preferable in all possible scientific and philosophical contexts. On the other hand, current debates about natural kinds (e.g. Magnus 2012; Franklin-Hall 2015; Slater 2014) suggest that the issue is far from settled and there is little reason to make my case for conceptual relativity dependent on some unnecessarily controversial theoretical proposal.

One way or another, the core result of this section remains the same. Indeed, conceptual relativity is incompatible with the idea that scientific kinds are discovered independently of our contingent interests. However, that does not mean that conceptual relativity reduces scientific kinds to mere conventions or inventions. On the contrary, conceptual relativity is entirely compatible with the broadly realist claim that scientific kinds typically reflect discoveries about property clusters and more generally about a reality that is largely independent from our interests and conceptualizations.[11] Instead, conceptual relativity is based on the assumption that these empirical discoveries underdetermine our scientific ontologies and do not lead to one unambiguous system of interest-independent natural kinds that "carve nature at its joints".

---

[11] It may be helpful to formulate this issue independently of the notion of natural kinds. Conceptual relativity (just as constructivism and other controversial philosophical positions) builds on the idea of multiple legitimate ways of conceptualizing reality. However, this idea seems to neglect the non-conceptualized structure of reality and one may object that it reduces reality to an "Amorphous Lump" (Eklund 2008) which we carve up in whatever way we want. Reality seems to become a silent supporting actor with no other role than ensuring philosophers that they are realists in some minimal sense. As Latour (2003) puts it: "one should add the comical role of being-there-just-to-prove-that-one-is-not-an-idealist role invented by Kant and replayed over and over again by philosophers all the way to David Bloor: things are there but play no role except that of mute guardians holding the sign 'We deny that we deny the existence of an outside reality'." A moderate account of natural kinds that acknowledges the importance of empirically discovered property clusters and of projectability addresses this worry by insisting that ontologies are shaped by interest-dependent conceptual choices but grounded in the empirically discovered structure of reality. However, this point does not necessarily have to be made in the "tradition of natural kinds" (Hacking 1991) and Latour's development from social constructivism to actor-network theory (Latour 2005, cf. Elder-Vass 2008) provides one alternative framework from a different philosophical tradition.

4.5   Realism and Existential Relativity

In the last section, I argued that conceptual relativity is entirely compatible with a moderate notion of natural kinds. Conceptual relativity may appear unacceptable to many contemporary metaphysicians, but its implications are not nearly as radical as they may appear at first. Although conceptual relativity contradicts an overly strong notion of natural kinds, there is no reason to believe that it is in conflict with the broadly realist idea that scientific kinds reflect a reality that is largely independent from our conceptualizations.

I expect that many philosophers will remain unconvinced by my claim that conceptual relativity is compatible with a satisfying realism and will object that conceptual relativity implies the radical and counterintuitive claim that truth values of scientific statements are relative to our conceptual decisions. This chapter is full of examples of this kind of relativity: "there are exactly three objects in Putnam's universe with three individuals" is true relative to mereological nihilism; "There exists a species to which *Quercus macrocarpa* and *Quercus bicolor* belong" is true relative to the biological species concept; "Some of my memories are realized by external media" is true relative to an externalist account of cognition; "Anna is more intelligent than Paul" is true relative to a specific intelligence concept; and so on. The relativity of truth values does not imply that any conceptual decision is as good as any other. There can be important pragmatic reasons to favor conceptual decisions in specific contexts, but many philosophers will worry that pragmatic considerations will not save us from a highly implausible anti-realism.

This worry appears even more pressing if we formulate it in terms of existence instead of truth values. According to conceptual relativists, not only truth values but also existence is relative to conceptual decisions: whether composed objects exist depends on conceptual decisions; whether a general intelligence exists depends on conceptual decisions; and so on. This kind of existential relativity seems to show that conceptual relativity comes at the price of an unacceptable anti-realism or even linguistic idealism. Again, this implication seems to be a central motivation of many metaphysicians who believe that conceptual relativity would mean that we have to capitulate to the "postmodern forces of darkness" (Sider 2012, 65). As Sider puts it: "The realist picture requires the 'ready made world' Goodman (1978) ridiculed; it requires a conception of the world as *really* being as physics says; it requires objectivity; it requires distinguished structure. To give up on structure's objectivity would be to concede far too much to those who view inquiry as being merely the investigation of our own minds" (2012, 65).

Although the worry that relative existence will lead to an unacceptable anti-realism is understandable, I think that it is ultimately misguided. Even if we accept that there are many legitimate ways of talking about the existence of objects or scientific kinds, this does not mean that existence reduces to linguistic choice. For example, there is a lot of empirical evidence that scientists do in fact talk differently about the existence of kinds. While some zoologists find cladistic species useful, others consider them "taxonomic artifacts" (Zachos et al. 2013). While some geneticists postulate continental genetic clusters, others employ different metrics and measures that lead to different genetic kinds (Kaplan and Winther 2012; Winther 2014). These varying ways of talking about the existence of scientific kinds, do not imply, however that existence reduces to linguistic choice. Once we specify our criteria, it depends on the empirically discovered structure of reality whether a certain kind exists.

If we turn from scientific practice to contemporary metaphysics, we find analogous considerations in debates about "quantifier variance" (Hirsch 2011) and "existential relativity" (Sosa 1999). More precisely, I want to argue the compatibility of realism and conceptual relativity is neatly summarized by Ernest Sosa's claim that "existence *relative* to a conceptual scheme is not existence *in virtue* of that conceptual scheme" (Sosa 1999, 134). Species pluralism, for example, implies that the existence of species is *relative* to a "conceptual scheme" in the sense that proponents of different species concepts answer the question what species exist in different ways. However, this does not mean that species exist *in virtue* of these conceptual choices. On the contrary, they exist in virtue of a biological reality that is clearly independent from our conceptualizations.

Consider the example of the North American Oak (e.g. Van Valen 1975). *Quercus macrocarpa* (Bur Oak) and *Quercus bicolor* (Swamp White Oak) are often considered two distinct species despite common hybridizations between both populations. Proponents of the ecological species concept can justify this distinction by pointing out that *Quercus macrocarpa* and *Quercus bicolor* inhibit distinct ecological niches while proponents of the biological species concept will insist that both populations belong to the same species due to hybridizations between them. Examples such as *Quercus macrocarpa* and *Quercus bicolor* can to illustrate the claim that existence relative to a conceptual scheme is not existence in virtue of that conceptual scheme. If we accept species pluralism, the existence of two different species *Quercus macrocarpa* and *Quercus bicolor* is indeed relative to the ecological species concept. However, these species do not exist in virtue of the ecological species concept but rather in virtue of ecological phenomena that are independent of our conceptual choices. As soon as we have chosen to work with a specific account of species, the question what species exist has an objective answer that entirely depends on the structure of the biological realm.

The same is true if we consider other examples of conceptual relativity in the empirical sciences. For example, I have argued that the existence of extended cognitive processes is relative to conceptual choices. Cognitive neuroscientists who are primarily concerned with internal mechanisms of the brain have good reasons to prefer an internalist account of cognition while psychologists who work on issues such as problem solving in complex environments have good reasons to prefer an externalist framework. This interest-relativity of cognitive ontologies leads to an interest-relativity of existence statements: whether it is correct to say that extended cognitive processes exist depends on our choice of a conceptual framework which again depends on our explanatory interests. However, this kind of existential relativity is compatible with a realist attitude according to which cognitive processes exits *in virtue* of a reality that is largely independent of our conceptualizations. Internalists and externalists recognize different patterns and therefore opt for different ontologies.

Finally, it is attractive to extend this moderately realist framework from the empirical sciences to philosophical ontologies. Recall the case of Putnam's universe with three individuals (x1, $x2$, x3). If we accept Putnam's account of conceptual relativity, we also have to accept that existence of composed objects, such as (x1 + $x2$) or (x1 + $x2$ + x3), is relative to our conceptual decisions. However, there is still a clear sense in which these objects exist *in virtue* of the individuals x1, $x2$, x3 and not in virtue of our conceptual choices. Once we have decided what ontology we use, existence depends on the world and not on us. If we choose the conceptual framework of mereological universalism, there exist exactly seven objects in Putnam's universe and it would be wrong to claim that there exist three or eight objects. If we choose the conceptual framework of mereological nihilism, there exist exactly three objects in Putnam's universe and it would be wrong to claim that there exist seven or two objects. And, of course, these implications are due to the structure of the universe (or, in this case, Putnam's toy universe).

To sum up, the distinction between "existence relative to" and "existence in virtue of" clarifies the relation between realism and conceptual relativity. Indeed, conceptual relativity implies that entities such as species, cognitive processes, intelligence(s), and composed objects exist relative to conceptual choices. This kind of existential relativity contradicts a strong notion of interest-independent natural kinds and the ideal of exactly one fundamental scientific ontology. However, existential relativity does not contradict more moderate variants of realism according to which entities exist in virtue of a reality that is largely independent of our conceptualizations. Anyone who claims that conceptual relativity leads to some unacceptable form of anti-realism or relativism would therefore have to show that this kind of moderate realism is actually not satisfying or even not coherent.

4.6 Reconsidering the Dialectical Situation

In the previous sections, I argued that conceptual relativity is ubiquitous in scientific practice and leads to ontological pluralism. Different explanatory interests in science require different frameworks that imply the existence of different scientific entities such as biological kinds, cognitive processes, intelligences, and so on. I have argued that all of this is largely unproblematic because it is compatible with a moderate realism regarding natural kinds and with the claim that scientific entities exist in virtue of a reality that is largely independent of our conceptualizations.

In the beginning of this chapter, I suggested that this plurality of scientific ontologies leads to a positive and a negative challenge of the ideal of exactly one fundamental ontology. We can now specify both challenges. The positive lesson is that ontological pluralism is not nearly as radical and implausible as many metaphysicians think it is. On the contrary, we not only have plenty evidence that scientists do in fact often rely of different ontologies but also good reasons to think that this conceptual relativity is perfectly compatible with a moderate realism. It is also not difficult to see how analogous points can be made with respect to philosophical ontologies. For example, I have argued that the existence of different tiger species is relative to the phylogenetic species concept but in virtue of phylogenetic relations that are independent of our conceptualization. In analogy: The existence of seven objects in Putnam's universe with three individuals is relative to the universalist framework but it is in virtue of the three individuals. In both cases, it is simply a mistake to assume that the rejection of one fundamental ontology will lead to an unacceptably strong anti-realism or relativism.

Beyond these positive remarks, however, there is also a negative challenge. If ontological pluralism is ubiquitous in the empirical sciences, why should we insist on the ideal of one fundamental ontology in philosophy? Or, to put it differently: if researchers in successful disciplines such as zoology or genetics talk about the existence of entities in a variety of ways, why is "quantifer variance" (Hirsch 2011) perceived as such a radical and implausible position in philosophy? Of course, it is not logically inconsistent to insist on the ideal of exactly one fundamental ontology in philosophy despite countless examples of conceptual relativity in the empirical sciences. For example, one could argue that empirical sciences such as zoology, genetics, or cognitive science are simply irrelevant for metaphysical considerations about the fundamental structure of reality and that philosophers who are interested in the question what entities fundamentally exist are therefore justified to ignore ontological pluralism in the life sciences.[12]

However, there are at least three related difficulties. First, the idea of one fundamental ontology seems to lose a lot of its philosophical relevance if it is completely detached from the reality of scientific practice. For example, let us assume the truth of some revisionary philosophical ontology that rejects the existence of composed objects (and/or vague objects, sets, properties, identity over time, and so on). Given this interpretation, both scientific and ordinary ontologies are terrible guides to the structure of reality as there are no genes, species, cognitive states, chairs, books, and so on. Should scientists or ordinary people care about such a fundamental philosophical ontology and revise their own ontologies? It seems obvious that the answer is no and even many revisionist metaphysicians attempt to develop compatibilist strategies that keep ontological practice intact in ordinary and scientific contexts (van Inwagen 1990; Contessa 2014 cf. Uzquiano 2004; Korman 2009). However, it then becomes very unclear how a fundamental ontology should be relevant for anyone beyond the esoteric circle of analytic metaphysicians.

Second, there is not only a problem of relevance but also of plausibility. Given that ontological pluralism seems to be common in scientific practice, we would indeed need very good reasons to stick to the ideal of one fundamental ontology in philosophy. I have suggested in the last sections that fear of anti-realism and relativism is not a good reason because conceptual relativity does not have especially radical implications. Some philosophers may prefer an even stronger realism but it

---

[12]This does not mean that analogous arguments are hard to find in the physical and chemical sciences. See, for example, Slater (2005) for a pluralist interpretation of chemical kinds, Ruphy (2010) for a pluralist interpretation of astrophysical kinds, and Atmanspacher and Primas (2003) for conceptual relativity in quantum physics.

is not clear how this preference would translate into a philosophical argument. Of course, there may be other reasons to stick to the ideal of one fundamental ontology in philosophy. In the next section, I will discuss the worry that conceptual relativity is self-defeating because it does not leave any room for substantial philosophical disputes. One way or another, we will need a very strong motivation for the claim that metaphysicians can have what is unavailable to scientists: exactly one fundamental ontology.

Finally, conceptual relativity in science directly challenges the ideal of one fundamental philosophical ontology under the assumption what I called "naturalism of scientific practice" in the introduction. Recall that I contrasted this naturalist methodology with a "naturalism of placement problems" that starts with a certain metaphysical picture such as physicalism. I suggested that the goal of a naturalist methodology should not be to validate a presupposed metaphysical picture but that metaphysics should adapt to the reality of scientific practice. A proponent of the ideal of one philosophical ontology cannot accept this kind of naturalism and ontological pluralism in scientific practice.

In fact, a philosophical debate about ontologies that is completely detached from scientific practice requires a very strong methodological anti-naturalism that also creates tension with common positions in analytic metaphysics. Consider, for example, Quine's claim that the question what exists is answered by our best scientific theories (Quine 1948). If we accept a pluralism of equally correct scientific theories that provide different answers the question of what exists, this leads us straight to conceptual relativity. Another Quinean thought that can be used to stress the relevance of scientific ontologies for discussions of philosophical ontologies is the often assumed continuity of philosophy with science. Often, this continuity is postulated with respect to both the topics and the methods of philosophy. Philosophical existence questions cannot be understood as completely separated from scientific existence questions and philosophical methods in answering existence questions can neither be understood as completely separated from scientific methods in answering existence questions. If conceptual relativity turns out to be ubiquitous in science, we would therefore need at least very good reasons to stick with the strong ideal of exactly one fundamental ontology in philosophy.

References

Adams, Frederick, and Kenneth Aizawa. 2011. *The Bounds of Cognition*. Malden, MA: Blackwell Publishing.

Agapow, Paul-Michael, et al. 2014. The impact of species concept on biodiversity studies. *The quarterly review of biology* 79 (2): 161–179.

Andreasen, Robin. 1998. A New Perspective on the Race Debate. *The British Journal for the Philosophy of Science* 49 (2): 199–225.

Anderson, John R. 2000. *Cognitive Psychology and Its Implications*. San Francisco: W.H. Freeman

Atmanspacher, Harald, and Hans Primas. 2003. Epistemic and Ontic Quantum Realities. In *Time, Quantum and Information,* eds. Lutz Castell and Otfried Ischebeck, 301–321. Berlin: Springer.

Avise, John C. 2000. Cladists in Wonderland. *Evolution* 54 (5): 1828–1832.

Baars, Bernard J. 1986. *The Cognitive Revolution in Psychology*. New York: Guilford Press.

Baddeley, Alan. 2007. *Working Memory, Thought, and Action*. Oxford: Oxford University Press.

Bapteste, Eric, and Dupré, John. 2013. Towards a Processual Microbial Ontology. *Biology & Philosophy* 28(2): 379–404.

Bapteste, Eric, and Boucher, Yan. 2009. Epistemological impacts of horizontal gene transfer on classification in microbiology. *Methods in Molecular Biology* 532: 55–72.

Barker, Matthew, and Velasco, Joel. 2013. Deep Conventionalism about Evolutionary Groups. *Philosophy of Science* 80(5): 971–982.

Boyd, Richard. 1999. Homeostasis, Species, and Higher Taxa. In *Species: New Interdisciplinary Essays*, ed. Robert A Wilson, 141–85. Cambridge, Mass.: MIT Press.

Boyd, Richard. 2010. Homeostasis, higher taxa, and monophyly. *Philosophy of Science*, 77(5): 686–701.

Brigandt, Ingo. 2011. Natural Kinds and Concepts: A Pragmatist and Methodologically Naturalistic Account. In *Pragmatism, Science and Naturalism*, eds. Jonathan Knowles and Henrik Rydenfelt, 171–196. Frankfurt am Main: Peter Lang.

Chen, Jie-Qi. 2004. Theory of Multiple Intelligences: Is It a Scientific Theory? *The Teachers College Record* 106 (1): 17–23.

Clark, Andy, and David Chalmers. 1998. The Extended Mind. *Analysis* 58: 7–19.

Clarke, Ellen. 2010. The Problem of Biological Individuality. *Biological Theory* 5 (4): 312–325.

Contessa, Gabriele. 2014. One's a Crowd: Mereological Nihilism without Ordinary-Object Eliminativism. *Analytic Philosophy* 55(2): 199–221.

Cracraft, Joel. 1983. Species Concepts and Speciation Analysis. *Current Ornithology* 17: 159–87.

Craver, Carl F. "Mechanisms and natural kinds." *Philosophical Psychology* 22.5 (2009): 575–594.

De Queiroz, Kevin. 1998. *The General Lineage Concept of Species, Species Criteria, and the Process of Speciation*. Oxford: Oxford University Press.

Devitt, Michael. 2011. Natural Kinds and Biological Realisms. *Carving Nature at Its Joints: Natural Kinds in Metaphysics and Science*, eds. Joseph Keim Campbell and Michael O'Rourke; Matthew Slater, 155–170. Cambrdige, Mass.: MIT Press.

Dupré, John. 2002. Is 'Natural Kind' a Natural Kind Term? *The Monist* 85 (1): 29–49.

Edwards, AWF. 2003. Human Genetic Diversity: Lewontin's Fallacy. *BioEssays* 25 (8): 798–801.

Eklund, Matti. 2008. The picture of reality as an amorphous lump. In *Contemporary Debates in Metaphysics*, ed. Ted Sider, John Hawthorne and Dean W Zimmerman, 382–396. Malden, MA: Blackwell Publishing.

Elder-Vass, Dave. 2008. Searching for realism, structure and agency in Actor-Network Theory. *The British Journal of Sociology* 59 (3): 455–473.

Ereshefsky, Marc. 2001. *The Poverty of the Linnaean Hierarchy: A Philosophical Study of Biological Taxonomy*. Cambridge: Cambridge University Press.

Ereshefsky, Marc and Thomas Reydon. 2015. Scientific Kinds. *Philosophical Studies* 172 (4): 969–986.

Foucault, Michel 1966. *Les mots et les choses: Une archéologie des sciences humaines*. Paris: Gallimard.

Franklin-Hall, Laura. 2015. Natural Kinds as Categorical Bottlenecks. *Philosophical Studies* 172: 925–948.

Gardner, Howard. 1985. *Frames of Mind: The Theory of Multiple Intelligences*. New York: Basic Books.

Glasgow, Joshua. 2008. *A Theory of Race*. New York: Routledge.

Goodman, Nelson. 1978. *Ways of Worldmaking*. Indianapolis: Hackett Publishing.

Gould, Stephen Jay. 1996. *The Mismeasure of Man*. New York: WW Norton & Company.

Gray, Wayne D., and Wai-Tat Fu. 2004. Soft Constraints in Interactive Behavior: The Case of Ignoring Perfect Knowledge In-the-World for Imperfect Knowledge In-the-Head. *Cognitive Science* 28 (3): 359–82.

Haber, Matt and Jay Odenbaugh. 2009. The Edges and Boundaries of Biological Objects. *Biological Theory* 4 (3): 219–224.

Hacking, Ian. 1991. A tradition of natural kinds. *Philosophical Studies* 61 (1): 109–126.

Hacking, Ian. 2004. *Historical Ontology.* Cambridge, Mass.: Harvard University Press.

Hacking, Ian. 2007. Natural kinds: rosy dawn, scholastic twilight, *Royal Institute of Philosophy Supplement,* 61, 203–239.

Häggqvist, Sören. 2005. Kinds, projectibility and explanation. *Croatian journal of philosophy* 13: 71–87.

Herrnstein, Richard J., and Charles Murray. 1994. *Bell Curve: Intelligence and Class Structure in American Life.* New York : Free Press.

Hirsch, Eli. 2011. *Quantifier Variance and Realism: Essays in Metaontology.* Oxford: Oxford University Press.

Hochman, Adam. 2013. Against the New Racial Naturalism. *Journal of Philosophy* 110: 331–351.

Horn, John L., and John J. McArdle. 2007. Understanding Human Intelligence Since Spearman. In *Factor Analysis at 100: Historical Developments and Future Directions,* eds. Robert Cudeck and Robert C MacCallum, 205–47. Mahwah: Lawerence Erlbaum.

Hunt, Earl. 1985. Verbal Ability. In *Human Abilities,* eds. Robert Sternberg, 31–58, New York: Freeman.

Van Inwagen, Peter. 1990. *Material Beings.* Cornell: Cornell University Press.

Irmak, Nurbay. 2014. Purpose-Relativity and Ontology. Dissertation at the University of Miami. http://scholarlyrepository.miami.edu/oa_dissertations/1164/

Jackson, Peter. 2001. Subspecies and Conservation. *Cat News,* 32.

Jensen, Arthur Robert. 1998. *The g factor: The science of mental ability.* Westport: Praeger.

Kaplan, Jonathan, and Winther, Rasmus. 2012. Prisoners of Abstraction? The Theory and Measure of Genetic Variation, and the Very Concept of 'Race'. *Biological Theory* 7 (4): 1–12.

Kaplan, Jonathan, and Winther, Rasmus. 2014. Realism, Antirealism, and Conventionalism about 'Race'. *Philosophy of Science* 81 (5): 1039–1052.

Khalidi, Muhammad Ali. 2013. *Natural Categories and Human Kinds: Classification in the Natural and Social Sciences.* Cambridge: Cambridge University Press.

Kitcher, Philip. 1984. Species. *Philosophy of Science* 51 (2): 308–33.

Kitcher, Philip. 1987. Ghostly whispers: Mayr, Ghiselin, and the Philosopheron the ontological status of species. *Biology and Philosophy* 2(2): 184–192.

Kitcher, Philip. 2008. Carnap and the Caterpillar, *Philosophical Topics,* 36 (1): 111–127.

Korman, Daniel Z. 2009. Eliminativism and the Challenge from Folk Belief. *Noûs* 43 (2): 242–264.

Leonelli, Sabina. 2013. Classificatory theory in biology. *Biological Theory* 7 (4): 338–345.

Lewis, David. 1983. New work for a theory of universals. *Australasian Journal of Philosophy,* 61 (4), 343–377.

LaPorte, Joseph. 2004. *Natural Kinds and Conceptual Change.* Cambridge: Cambridge University Press.

Latour, Bruno. 2003 The Promises of Constructivism *Chasing technoscience: Matrix for materiality,* eds. Don Ihde et al., 27–46. Bloomington: Indiana University Press.

Latour, Bruno. 2005. *Reassembling the Social-An Introduction to Actor-Network-Theory.* Oxford: Oxford University Press.

Longino, Helen E. 2013. *Studying human behavior: how scientists investigate aggression and sexuality.* Chicago: University of Chicago Press.

Love, Alan. 2009. Typology reconfigured: from the metaphysics of essentialism to the epistemology of representation. *Acta Biotheoretica* 57 (1–2): 51–75.

Ludwig, David. 2013. Extended Cognition in Science Communication. *Public Understanding of Science.*

Ludwig, David. 2014a. Hysteria, Race, and Phlogiston. A Model of Ontological Elimination in the Human Sciences. *Studies in History and Philosophy of Science Part C: Studies in History and Philosophy of Biological and Biomedical Sciences* 45: 67–77.

Ludwig, David. 2014b. Extended Cognition and the Explosion of Knowledge. *Philosophical Psychology*.

Ludwig, David. 2014c. Disagreement in Scientific Ontologies. *Journal for General Philosophy of Science*, 45 (1): 119–131.

Ludwig, David. 2015a. Against the New Metaphysics of Race. *Philosophy of Science, 82 (2): 244-265.*.

Ludwig, David. 2015b. Indigenous and Scientific Kinds . *The British Journal for the Philosophy of Science, first published online.*

MacLeod, Miles, and Thomas AC Reydon. 2013. Natural Kinds in Philosophy and in the Life Sciences: Scholastic Twilight or New Dawn?. *Biological Theory* 7 (2): 89–99.

Magnus, Paul D. 2012 *Scientific Enquiry and Natural Kinds: From Planets to Mallards*. New York: Palgrave Macmillan.

Magnus, Paul D. 2014. NK ≠ HPC. *The Philosophical Quarterly* 64 (256): 471–477.

Mayr, Ernst. 1973. *Animal Species and Evolution*. Cambridge, Mass: Belknap Press.

Meier, Rudolf and Rainer Willmann. 2000. The Hennigian Species Concept *Species Concepts and Phylogenetic Theory: A Debate*, eds. Quentin D Wheeler and Rudolf Meier, 30-43. New York: Columbia University Press.

Miller, George A. 1956. The Magical Number Seven, Plus or Minus Two: Some Limits on Our Capacity for Processing Information. *Psychological Review* 63 (2): 81.

Millikan, Ruth Garrett. 1999. Historical Kinds and the 'Special Sciences'. *Philosophical Studies* 95 (1): 45–65

O'Malley, Maureen and Dupré, John. 2005. Fundamental Issues in Systems Biology. *BioEssays*, 27 (12): 1270–1276.

Quicke, Donald L. J. 1993. *Principles and Techniques of Contemporary Taxonomy*. London: Blackie Academic & Professional.

Quine, Willard Van Orman. 1948. On What There Is. *Review of Metaphysics* 2 (5): 21—36.

Pöyhönen, Samuli. 2013. Explanatory Power of Extended Cognition. *Philosophical Psychology*.

Reydon, Thomas AC. 2005. On the nature of the species problem and the four meanings of 'species'. *Studies in History and Philosophy of Science Part C: Studies in History and Philosophy of Biological and Biomedical Sciences* 36 (1): 135–158.

Reydon, Thomas AC. 2009. Species in three and four dimensions. *Synthese* 164 (2): 161–184.

Rieppel, Olivier. 2005. Monophyly, Paraphyly, and Natural Kinds. *Biology and Philosophy* 20 (2–3): 465–487.

Rieppel, Olivier. 2013. Biological Individuals and Natural Kinds. *Biological Theory* 7 (2): 162–169.

Rowlands, Mark. 2003. *Externalism: Putting Mind and World Back Together Again*. Montreal: McGill-Queen's Press.

Rupert, Robert D. 2009. *Cognitive Systems and the Extended Mind*. Oxford: Oxford University Press.

Ruphy, Stéphanie. 2010. Are Stellar Kinds Natural Kinds? A Challenging Newcomer in the Monism / Pluralism and Realism / Antirealism Debates. *Philosophy of Science* 77 (5): 1109–20.

Sider, Ted. 2012. *Writing the Book of the World*. Oxford: Oxford University Press.

Slater, Mathew. 2005. Monism on the One Hand, Pluralism on the Other. *Philosophy of Science* 72 (1): 22–42.

Slater, Mathew. 2014. Natural kindness. *The British Journal for the Philosophy of Science*.

Sosa, Ernest. 1999. Existential Relativity. *Midwest Studies in Philosophy* 23 (1): 132–43.

Spearman, Charles. 1904. 'General Intelligence,' Objectively Determined and Measured. *The American Journal of Psychology* 15 (2): 201–92.

Spencer, Quayshawn (2014). A Radical Solution to the Race Problem. *Philosophy of Science* 81 (5): 1025–1038.

Stanford, P. Kyle. 1995. For pluralism and against realism about species. *Philosophy of Science* 62 (1): 70–91.

Thurstone, Louis Leon. 1938. *Primary Mental Abilities*. Chicago: University of Chicago Press.

Uzquiano, Gabriel. 2004. Plurals and simples. *The Monist* 87 (3): 429–451.

Van Valen, Leigh. 1976. Ecological Species, Multispecies, and Oaks. *Taxon* 25 (2): 233–39.

Waterhouse, Lynn. 2006. Multiple Intelligences, the Mozart Effect, and Emotional Intelligence: A Critical Review. *Educational Psychologist* 41 (4): 207–25.

Wilson, Robert A., and Matthew Barker. 2013. The biological notion of individual. *The Stanford Encyclopedia of Philosophy*, eds. Edward N. Zalta. http://plato.stanford.edu/entries/biology-individual/

Wilson, Robert, Matt Barker, and Ingo Brigandt. 2007. When Traditional Essentialism Fails: Biological Natural Kinds. *Philosophical Topics* 35: 189–215.

Winther, Rasmus. 2011. Part-Whole Science. *Synthese* 178 (3): 397–427.

Winther, Rasmus. 2014. The Genetic Reification of Race?: A Story of Two Mathematical Methods. *Critical Philosophy of Race* 2 (2): 204–223.

Winther, Rasmus and Kaplan, Jonathan. 2013. Ontologies and Politics of Biogenomic 'Race'.: *Theoria* 60 (136): 54–80.

Zachar, Peter. 2002. The practical kinds model as a pragmatist theory of classification, *Philosophy, Psychiatry, & Psychology* 9 (3): 219–227.

Zachos, Frank E., et al. 2013. Species Inflation and Taxonomic Artefacts—A Critical Comment on Recent Trends in Mammalian Classification. *Mammalian Biology-Zeitschrift für Säugetierkunde* 78 (1): 1–6.

Chapter 5
The Demarcation Problem of Conceptual Relativity

In the previous sections, I argued that conceptual relativity is common in the empirical sciences as there are often different but equally legitimate scientific answers to the question what entities exist. Given the biological species concept, it is correct to say that only one tiger species exists. Given the phylogenetic species concept, it is correct to say that several tiger species exist. As both accounts of species can be justified through different explanatory interests in biological research, we should accept a plurality of equally legitimate answers to the question what species exist. Furthermore, I have argued that conceptual relativity is entirely compatible with a moderate realism according to which entities exist relative to our conceptual decisions but still in virtue of a reality that is independent of our conceptualizations. Finally, I have suggested that examples from the empirical sciences lead to a negative and a positive challenge of the ideal of one fundamental ontology in philosophy: on the negative side, it is at least unclear why philosophers should aim for exactly one fundamental ontology given the ubiquity of conceptual relativity in the empirical sciences. On the positive side, conceptual relativity does not lead to some unacceptable philosophical radicalism but is rather compatible with a moderate realism.

Proponents of the ideal of exactly one fundamental ontology will object that this presentation understates the dangers of conceptual relativity. One potential problem is that conceptual relativity seems to lead to a strong deflationist attitude. In the case of the empirical sciences, I have suggested a deflationist account of debates about species, cognition, and intelligence by arguing for a plurality of different but equally legitimate ontologies. An extension of conceptual relativity to debates about philosophical ontologies also extends the deflationist attitude to issues such as the debate between nihilists and universalists. One legitimate worry is that conceptual relativity will lead to an excessive deflationism and maybe even to the self-defeating claim that *all* existence disputes can be dissolved by distinguishing between different ontologies.

One way of specifying this worry is based on current philosophical debates about verbal disputes (e.g. Chalmers 2011; Jackson 2013; Jenkins 2014). I have suggested

© Springer International Publishing Switzerland 2015
D. Ludwig, *A Pluralist Theory of the Mind*, European Studies
in Philosophy of Science 2, DOI 10.1007/978-3-319-22738-2_5

that many existence disputes in scientific and philosophical ontologies are verbal disputes in the sense that they do not arise from different beliefs about reality but from the deployment of different conceptual frameworks. However, not every existence dispute is a verbal dispute and a plausible account of conceptual relativity therefore has to leave room for substantive disputes about existence questions. It seems that conceptual relativists are obligated to draw a line between merely verbal and substantive existence disputes and have to explain why their deflationism does not affect *every* metaphysical and scientific existence dispute. This is what I will call the "demarcation problem of conceptual relativity." In the following sections, I will argue that the demarcation problem affects both proponents and critics of conceptual relativity and that many attractive answers to the demarcation problem actually favor conceptual relativity.

5.1   Verbal and Substantive Disputes

In order tackle the demarcation problem, it is important to notice that the question of whether or not a dispute is merely verbal can only be answered if its context is taken into account. Consider the following examples:

(20) There are dandelion populations in the tropics
(21) There are no dandelion populations in the tropics

(22) God exists
(23) God does not exist

(20) vs. (21) and (22) vs. (23) seem to be paradigmatic cases of substantive disputes: either there are dandelion populations in the tropics or there are no dandelion populations in the tropics; and either God exists or God does not exist. There does not seem to be anything verbal about these disputes. Things are, however, more complicated. Imagine two biologists debating whether or not there are dandelion populations in the tropics. At first, it seems that there is a perfectly substantive disagreement and that only one of the biologists can be right. However, there are different plausible interpretations of their claims. On the one hand, "dandelion" is often taken to refer to *Taraxacum officinale*, a species native to Eurasia and naturalized throughout other temperate regions. Given this interpretation, the claim that there are dandelion populations in the tropics is wrong. On the other hand, "dandelion" is often taken to refer to the genus *Taraxacum* which includes *Taraxacum officinale*, but also other species that are native to tropical regions. Given this interpretation, the claim that there are dandelion populations in the tropics is true. The moral of this example is that it depends on the context whether debates about questions such as (20) vs. (21) are merely verbal or not: if one biologist refers to *Taraxacum officinale* while the biologist refers to the genus *Taraxacum*, the debate turns out to be merely verbal. If both of them refer to *Taraxacum officinale* (or to the genus *Taraxacum*), the debate turns out to be substantive.

The situation is similar in the case of metaphysical debates such as (22) vs. (23). Consider an atheist debating with a person who claims that God exists. Although we will start with the assumption of a substantive disagreement, the situation becomes less clear when we learn that the proponent of (22) rejects the idea of a personal god and instead claims that "god is simply everything that exists." The dispute becomes even more suspicious when we learn that she considers herself as a "naturalist pantheist" who rejects immaterial souls and supernatural entities, and instead endorses a physicalism. In this situation, it seems obvious that the appearance of a substantive disagreement breaks down. The proponents of (22) and (23) do not disagree about the world; they have a merely verbal dispute about what "God exists" means.

Debates such as (20) vs. (21) and (22) vs. (23) are good examples for the demarcation problem. Intuitively it seems obvious that there are contexts in which the debates should be considered "merely verbal" and other contexts in which they are substantive. But how can we distinguish between them? The examples are helpful starting points in searching for a solution: in both cases, we are inclined to describe the disputes as "merely verbal" if they arise wholly in virtue of semantic differences. The debates are therefore merely verbal if both sides simply mean different things with "There are dandelion populations in the tropics" or "God exists."

The idea that we can identify merely verbal disputes by asking whether they arise wholly in virtue of semantic differences fits well with many contemporary accounts of verbal disputes. For example, David Manley argues that "a dispute is verbal just in case the speakers disagree because they semantically deviate from each other" (2009, 14). Another account of verbal disputes that presents semantic differences as decisive is suggested by Chalmers: "A dispute over S is (broadly) verbal when for some expression T in S, the parties disagree about the meaning of T, and the dispute over S arises wholly in virtue of this disagreement regarding T" (2011, 522). Given these accounts of merely verbal disputes, it might seem that we have a handy criterion to solve the demarcation problem:

(DC1) A dispute is merely verbal iff it arises wholly in virtue of semantic differences

(DC1) fits well the examples of merely verbal disputes discussed so far. In the case of the dandelion populations, the dispute between proponents of (20) and (21) arises wholly in virtue of semantic differences concerning "dandelion population," if one party means *Taraxacum officinale* and the other one means the genus *Taraxacum*. However, it does not arise wholly in virtue of semantic differences if they both mean *Taraxacum officinale* or the genus *Taraxacum*.

Although (DC1) is an attractive proposal, it is not without problems. First, there is a subtle but important difference between "verbal disputes" and "*merely* verbal disputes." In the case of "merely verbal disputes," we are inclined to argue that the dispute is superficial and both sides are equally correct. However, not every verbal dispute is merely verbal in this sense. Consider the following example about how to translate the German word "sympathisch:"

(24) "sympathisch" means "likeable"
(25) "sympathisch" means "sympathetic"

Clearly, this is a verbal dispute in the sense that it arises wholly in virtue of semantic differences, but it still seems wrong to lump it together with "merely verbal disputes." Proponents of (24) and (25) do not debate a superficial question and they are not equally correct; the proponent of (25) is simply wrong in assuming that "sympathisch" should be translated as "sympathetic." Therefore, (DC1) is not sufficient to identify *merely* verbal disputes.

It might be possible to solve this problem by adding another condition to (DC1). For example, one might require that a dispute is *merely* verbal only if the parties are engaged in a *prima facie* non-verbal dispute which arises wholly in virtue of semantic differences. In the case of (24) vs. (25) both parties are fully aware that they are debating a semantic issue and they intend to debate a semantic issue. Merely verbal disputes have a different structure; the parties intend to discuss a non-semantic question but fail to do so (cf. Jenkins 2014).

Even if an additional condition along these lines would solve the problem posed by disputes such as (24) vs. (25), there would remain another problem for the use of (DC1) as an answer to the demarcation problem: if people disagree on the question whether a debate is merely verbal, they will probably also disagree on the question of whether it arises wholly in virtue of semantic differences. This is well illustrated by the debate between mereological nihilists and universalists. A conceptual relativist would argue that disputes such as (12) vs. (13) arise in virtue of semantic differences, because we can talk about the existence of objects in different ways. In contrast, ontological realists will insist that there is only one correct interpretation of the existential quantifier and that disputes about (12) vs. (13) are therefore not in virtue of semantic differences. An application of (DC1) will not solve the debate because there will remain disagreement on the question of semantic differences.

5.2   Interpretive Charity as an Answer to Demarcation Problem

(DC1) does not solve but reformulates the demarcation problem, so how can we ascertain whether a dispute arises wholly in virtue of semantic differences? Sometimes this question is easy to answer. Consider cases in which the disagreement is about sentences that are definitional equivalents to sentences that both parties agree on. A highly simplified version of the species debate can provide an example. Recall the debate between a proponent of the ecological species concept and a proponent of the biological species concept in the case of *Quercus macrocarpa* and *Quercus bicolor*. The problem is that they can produce fertile offspring but do not inhabit the same ecological niche. Therefore, proponents of a biological and ecological species concept disagree with respect to question whether *Quercus macrocarpa* and *Quercus bicolor* belong to the same species.

Let us assume for the sake of the argument (again, this is a simplification) that the biologist who uses the ecological species concept takes "belong to the same species" to be definitional equivalent to "inhabit the same niche" while the biologist who uses the biological species concept takes "belong to the same species" to be definitional equivalent to "be able to produce fertile offspring." This allows us to reformulate (14) and (15):

(14) There exists a species to which *Quercus macrocarpa* and *Quercus bicolor* belong is a definitional equivalent to

(14)''' *Quercus macrocarpa* and *Quercus bicolor* can produce fertile offspring

(15) There exists no species to which *Quercus macrocarpa* and *Quercus bicolor* belong is a definitional equivalent to

(15)''' *Quercus macrocarpa* and *Quercus bicolor* do not inhabit the same niche

If both biologists agree on (14)''' and (15)''', understood as definitional equivalents of (14) and (15), then we can conclude that the debate arises wholly in virtue of semantic differences concerning the question of what it means "to belong to the same species." The example suggests a different answer to the demarcation problem:

(DC2) A dispute is merely verbal iff the disputed claims are definitional equivalent to sentences on which both parties agree

(DC2) may formulate a sufficient condition for merely verbal disputes, but the condition is too strong since we do not always have a definitional equivalent sentence available. Chalmers (2011) illustrates this problem with Ishmael's famous speech in Melville's Moby Dick:

> Be it known that, waiving all argument, I take the good old fashioned ground that the whale is a fish, and call upon holy Jonah to back me. This fundamental thing settled, the next point is, in what internal respect does the whale differ from other fish. Above, Linnaeus has given you those items. But in brief, they are these: lungs and warm blood; whereas, all other fish are lungless and cold blooded (1851/2013, 35).

The example is similar to the *prima facie* contradiction between the ecological and the biological species concept. If (14) vs. (15) turns out to be merely verbal, the disagreement between Ishmael and Linnaeus should be merely verbal too. However, Ishmael and Linnaeus do not need to have a precise definition on what it means to "be a fish." And if there are no definitional equivalent sentences, (DC2) suggests that the debate is not merely verbal.

Definitional equivalence might be too strong for a demarcation criterion, but I still think we are on the right track. What I take to be an important insight of (DC2) is that the question of whether we should consider a dispute as merely verbal depends on how we should interpret the disputed claims. In the case of Ishmael and Linnaeus, there might be no definitional equivalent sentence, but both parties could still interpret each other as speaking the truth in their own language. For Ishmael, "to be a fish" roughly means to look and to behave like a fish. For Linnaeus, "to be a fish" means to share certain morphological properties with other fish. Given this clarification, they can interpret each other as claiming:

(26) Whales look and behave like fish
(27) Whales do not share certain morphological properties with fish

If Linnaeus would interpret Ishmael to claim (26), while Ishmael would interpret Linnaeus to claim (27), there would be no disagreement left as Linnaeus and Ishmael agree on (26) and (27). The example suggests a further answer to the demarcation problem

(DC3) A dispute is merely verbal iff both sides can interpret each other as speaking the truth in their own language

(DC3) largely rests on Hirsch's account of verbal disputes, according to which a controversy is merely verbal "if the following condition is satisfied: Each side can plausibly interpret the other side as speaking a language in which the latter's asserted sentences are true."[1] But how can we determine whether this condition is satisfied? Whether or not the condition is satisfied cannot depend on both sides *actually* interpreting each other as speaking the truth in their own language. For example, Ishmael does not interpret Linnaeus as speaking the truth in his own language, but the disagreement is still merely verbal. Therefore, (DC3) must be understood as a normative criterion. The demarcation criterion is not whether both sides *will* interpret each other interpret each other as speaking the truth in their own language but whether they *should* do so. But when should both parties interpret each other as speaking the truth in their own language?

One way of clarifying (DC3) is to invoke the idea of charitable interpretation: both sides should interpret each other as speaking the truth in their own language iff there is a charitable interpretation according to which they are speaking the truth in their own language:

(DC4) A dispute is merely verbal iff there is a charitable interpretation according to which both parties should interpret each other as speaking the truth in their own language

(DC4) is only a clarification of (DC3) if we specify what counts as a "charitable interpretation". Although the principle of charity was introduced by Neil L. Wilson (1959, 532) to contemporary philosophy, it is most commonly associated with Donald Davidson's theory of radical interpretation. According to Davidson, an interpreter of a different language has to use the principle of charity by maximizing both agreement and consistency:

---

[1] Hirsch (2008, 231). See Hirsch (2011) for a collection of his papers on metaontology and verbal disputes.

Charity in interpreting the words and thoughts of others is unavoidable … just as we maximize *agreement*, or risk not making sense of what the alien is talking about, so we must maximize the self-consistency we attribute to him, on pain of not understanding *him*. (Davidson 1984, 27)

According to Davidson, "maximizing agreements" means that we should take another person to be right by the interpreter's lights as often as possible. Of course, the appeal to "maximizing agreement" can only be one part of the story. The principle of charity would be absurd, if we would always assume that the other person is speaking the truth in her own language and if we would classify every disagreement as merely verbal. There must be room for substantive disagreement.

How do we find substantive disagreement given that the principle of charity advises us to maximize agreement? The general answer is that we do not only have to maximize agreement; we also have to maximize consistency. Consider a simple case such as Anna and Paul having a dispute about the question of whether there is still a beer in Anna's fridge. Anna claims that there is still a beer in the fridge, while Paul insists that there is no beer in the fridge. Let us assume that Anna and Paul actually have a substantive disagreement and that Paul falsely believes that there is no beer in the fridge. If there were a charitable interpretation according to which Anna and Paul should interpret each other as speaking the truth in each others languages, (DC4) would have to be rejected.

Does (DC4) suggest that both are speaking the truth in their own language? One way of arriving at this conclusion is to assume that they should interpret each other as speaking about different fridges. If Anna would interpret Paul as saying that there is no beer in *his* fridge, she could understand his claim as being consistent and true. Does this mean, then, that both requirements of the principle of charity (to maximize consistency and maximize truth) are met, and that Anna should interpret Paul as speaking about a different fridge even if they are actually speaking about the same fridge? The obvious answer is that a charitable interpretation has to go beyond single sentences and that Anna can easily figure out whether Paul is speaking about a different fridge. For example, she can ask him which fridge he means. If Paul says that he is also talking about Anna's fridge, Anna has every reason to believe that they are talking about the same fridge. Furthermore, she can go to her fridge and take a beer out. If Paul retracts his original statement, then they were talking about the same fridge. If he says something like "Oh, I meant my fridge!", then their dispute was obviously verbal because they referred to different fridges. Given the consistency constraint of the principle of charity, there are obviously many situations in which it is not legitimate to interpret Paul as referring to a different fridge and therefore speaking the truth in his own language.

One may object that we can still come up with a different charitable interpretation, according to which Paul is speaking the truth in his own language. For example, we could assume that Paul is a "beer snob" who does not consider light beers to be "real beers." When he sees a light beer he often says things like "That's not a real beer, that's water in a can!" If Paul uses the word "beer" in this idiosyncratic way, both Anna and Paul might be speaking the truth in their own language. However, it is easy to figure out whether this is the case. For example, Anna can simply ask Paul whether he's just trying to make the point that her beer is not "real beer." In addition, she can open the fridge, take out a beer, and see whether Paul retracts his statement.

So far, I have discussed two scenarios which might lead Anna and Paul to the assumption that their dispute is merely verbal. It is easy to verify or falsify these scenarios, however, and they therefore do not provide counter examples to (DC4). Of course, we can think of scenarios in which it will be considerably more complicated to answer the question of whether it is possible to interpret both sides as speaking the truth in their own language and my examples cannot count as a proof of the correctness of (DC4).[2]

Still, (DC4) is clearly more promising than (DC1) – (DC3) as an answer to the demarcation problem and it is not difficult to imagine its application to debates about scientific disputes. For example, imagine two zoologists conducting fieldwork on a small Indonesian island off the coast of Sulawesi. One of their goals is to compile a complete list of bat species that are native to the island. They decide to conduct fieldwork independently from each other, to collect specimens, and to compare their lists after a few weeks. As it turns out, one list includes 25 bat species while other only includes 23 bat species.

The situation is compatible with both a verbal or a substantive disagreement between the zoologists. One the one hand, the first zoologist may have been more successful in her observation of rare bat species and the second zoologist may be simply wrong in her assumption of 23 bat species. On the other hand, the disagreement may be verbal as both zoologists use established but slightly different criteria for species membership. In this case, both sides both sides would speak "the truth in their own language", as Hirsch puts it.

(DC4) suggests that the decision between both scenarios should be based on a charitable interpretation that maximizes both agreement and consistency. In the spirit of maximizing agreement, the zoologists may check whether they are simply using different taxonomies. This assumption can be easily verified or falsified. First, the zoologists can simply discuss the criteria they were using in distinguishing between bat species. If there remain any doubts, they may also have a look at the empirical evidence and compare their specimen collections. For example, the second zoologist may retract her statement if the collection of the first zoologist includes specimens that she didn't find. The application (DC4) therefore does not imply that there substantial disagreement disappears through the attempt of maximizing agreement. On the contrary, (DC4) seems to provide a helpful tool of distinguishing verbal and substantive disputes both in ordinary and scientific contexts.

---

[2] Compare, for example, Warren's (2014) discussion on the "collapse argument" and the exchange between Jackson (2013) and Hirsch (2013) on the question whether ontological debates will include "unrevisable" sentences given Hirsch's deflationism.

5.3 Turning the Demarcation Problem Upside Down

The demarcation problem constitutes an important challenge for proponents of conceptual relativity. If conceptual relativists claim that some existence disputes are verbal because both sides deploy equally correct conceptual frameworks, they have to find a way of distinguishing them from substantive disputes. Otherwise they run the risk of claiming that every existence dispute is a verbal dispute. This result would arguably imply a radical relativism that should be considered a *reductio* of conceptual relativity.

In this section, I want to argue that the demarcation problem does not only affect proponents but also critics of conceptual relativity. Critics of conceptual relativity who insist on the non-verbal character of ontological disputes in science and philosophy also have to find a way of distinguishing them from verbal disputes. Without an answer to the demarcation problem, proponents of the ideal of one fundamental ontology would not be able to identify any disputes as verbal disputes. In other words: while the conceptual relativist has to leave room for substantive non-verbal disputes, her critic has to leave room for non-substantive verbal disputes.

Demarcation criteria such as (DC4), however, *prima facie* support my claim that conceptual relativity is common in science. Consider, for example, my discussion of externalism in cognitive science. (DC4) suggests that a dispute between externalists and internalists is verbal, if there is a charitable interpretation according to which both parties should interpret each other as speaking the truth in their own language. Furthermore, it seem almost obvious that such interpretation is available. When an externalist claims that extended cognition exists, a charitable internalist can reinterpret this claim as referring to both internal cognitive processes and external non-cognitive media that often play an important role in cognitive routines. When an internalist claims that extended cognitive processes no not exist, a charitable externalist would reinterpret this claim as referring to the biological part of cognition.[3]

On a more general level, the role of explanatory interests in scientific ontologies supports the assumption that (DC4) will often lead to conceptual relativity in the empirical projects. My discussion of species, extended cognition, and intelligence as well as natural kinds in general suggests that scientists work with different ontologies *because* they find different kinds meaningful in their research contexts. Given that different ontologies are well motivated in scientific practice, it is easy to maximize not only agreement but also to maintain consistency by interpreting the other side as speaking the truth in their own language.

Finally, charity-based criteria such as (DC4) do not only seem to imply conceptual relativity in the case of scientific ontologies but also in the case of traditional philosophical examples such as the dispute between nihilists and universalists.

---

[3] This does not mean that the debate between externalists and internalists are pointless or *merely* verbal in a negative sense. On the contrary, there are often very good pragmatic reasons to consider debates about ontologies in science important.

Recall the dispute between nihilists and universalists in the case of Putnam's universe with three individuals:

(12) There are exactly three objects
(13) There are exactly seven objects

Should proponents of (12) and (13) interpret each other as speaking the truth in their own language, given the criteria of a charitable interpretation? The principle of charity advises us to maximize agreement without making the other person's believe system inconsistent. Therefore, proponents of (12) and (13) should interpret each other as speaking the truth in their own language if and only if that does not make the other person's believe system inconsistent.

It is helpful to contrast this ontological dispute with uncontroversial examples of substantive disputes. Recall the case of Anna and Paul having a substantive dispute about whether there is still a beer in the fridge. We have seen that there are simple ways to verify or falsify whether Anna and Paul have a substantive disagreement. Most obviously, we can open the fridge and see whether one of them retracts the statement. To use another example, suppose that Anna and Paul disagree about how many objects exist in Putnam's universe with three individuals because they disagree how many individuals exist in Putnam's universe; Anna knows that there are three individuals while Paul falsely believes that there are seven individuals. Both Anna and Paul only count individuals as objects and therefore disagree about (12) and (13). If we try to maximize agreement by taking both Anna and Paul to speak the truth in their own language we will immediately create what Hirsch calls "cascading complications" (2005, 73): if we interpret the dispute as verbal, we create inconsistencies. We can deflate these inconsistencies by interpreting them as verbal, but this will create even more inconsistencies. Again, we can interpret them as verbal, but this will only take us into deeper interpretive trouble.

If, for example, we consider Anna's and Paul's dispute about (12) and (13) as being verbal, how do we explain their disagreement concerning the following statement?

(28) There are exactly three individuals
(29) There are exactly seven individuals

We could try to maintain consistency by arguing that the disagreement between (28) and (29) is also verbal, and that both Anna and Paul mean something different by "existence of individuals." However, this reinterpretation will create new and cascading interpretive problems. If we explain the disagreement over (28) and (29) by claiming they mean something different by "existence of individuals," then we face new problems in other contexts in which Anna and Paul actually agree on the number of individuals. Furthermore, we also run into trouble when Paul is presented with Putnam's universe with three individuals and retracts his statement by saying: "I was wrong. There are only three individuals and therefore only three objects."

Nothing like this will happen in the case of nihilists and universalists, since not only do they agree on the number of individuals, they also will not see any reason to retract their statements. If we present both of them with Putnam's universe, they will both feel assured in their claims that there are exactly three/seven objects. As the allegedly substantive dispute is disconnected from other substantive disagreements, we can use the principle of charity to argue that both sides speak the truth in their own language without running into risk of creating inconsistencies. If we accept (DC4), it seems their debate should be considered verbal.

A critic of conceptual relativity who endorses on the ideal of exactly one fundamental ontology might reply that this application of the principle of charity misses an important aspect: nihilists and universalists *insist* that they have a substantive disagreement. By claiming that they both speak the truth in their own language we imply that they are wrong about the very nature of their disagreement. Therefore, describing their dispute as verbal is everything but a *charitable* interpretation. Indeed, if we consider the debate between nihilists and universalists to be verbal, we also have to say that they are wrong in thinking that they debate a substantive question. However, this is not enough to show that the most charitable interpretation implies a substantive dispute. Recall the case of Ishmael and the whale: although Ishmael considers his disagreement with Linnaeus as substantive, we are inclined to say that the disagreement is verbal. Furthermore, Ishmael is stubborn. Even if we explain to him why this debate is verbal, he might still insist that his disagreement with Linnaeus is substantive and that whales *really* are fish. This will not, however, change anything about the fact that the most charitable interpretation considers both parties to be speaking the truth in their own language. A dispute does not become substantive simply because the contestants insist that it is substantive.

Given that the principle of charity seems to suggest that both nihilists and universalists speak the truth in their own language, a critic of conceptual relativity will probably reject (DC4) as an answer to the demarcation problem. And indeed, dedicated ontologists such as Sider (2009, 392) have argued that the appeal to charity is not enough, as some accounts are more eligible than others. In the case of nihilism and universalism, a charitable interpretation might suggest that both parties speak the truth in their own language. However, Sider argues that the decisive question is whether the nihilist's or the universialist's account is more eligible.

5.4 Joint Carving and Similarity

"Eligibility" is a highly ambiguous term and has many innocent interpretations. For example, we can consider one description more eligible than another because it better fits the language that we are already speaking. Consider the famous dialogue between Lewis Carroll's Alice and Humpty Dumpty:

"I don't know what you mean by 'glory,'" Alice said.
Humpty Dumpty smiled contemptuously. "Of course you don't—till I tell you. I meant 'there's a nice knock-down argument for you!'"
"But 'glory' doesn't mean 'a nice knock-down argument,'" Alice objected.
"When I use a word," Humpty Dumpty said, in a rather a scornful tone, "it means just what I choose it to mean—neither more nor less." (Carroll 1951, 189–190).

Of course, Alice and Humpty Dumpty can interpret each other as speaking the truth in their own language, but there is still an obvious sense in which Alice's concept of "glory" is more eligible. Alice speaks plain English while Humpty Dumpty speaks a pointless artificial language. The example suggests an interpretation of "eligibility" according to which one description is more eligible than others if it is closer to our ordinary language. Obviously, this is not what Sider means with "eligibility", since he is not interested in our ordinary language but a fundamental account of reality that will be radically different from our ordinary language.

There is another fairly innocent interpretation of "eligibility" that is based on pragmatic considerations. Imagine a person who rejects common biological taxonomies and instead defines taxa based on color. For example, she denies that a brown cat and a white cat belong to same species because they have different colors, and she insists that a brown cat and a brown cow belong to the same species because they have the same color. Obviously, her descriptions are less eligible than the descriptions of biologists. One reason why color-based taxa are less eligible than common biological taxa is that they are useless in research. There are no realistic contexts in which color-based species concepts would turn out to be useful for biologists, whereas common biological taxa have proven extraordinarily useful in scientific practice (cf. 4.4). Again, this interpretation of "eligibility" does not fit well the ideal of exactly one fundamental ontology. Usefulness is context-dependent, while critics of conceptual relativity want to defend the idea of one fundamental and therefore context-independent ontology.

How then shall we understand the appeal to eligibility? Sider answers this question in the following way: "Eligibility I understand as naturalness: a candidate meaning is more eligible if it 'carves nature at the joints.'" (Sider 2001, 198). Sider's account of eligibility as joint carving illustrates how critics of conceptual relativity can reject the appeal to interpretive charity. Even if the principle of charity suggests that both sides speak the truth in their own language, only one of the languages involves candidate meanings that carve nature at its joints. Charity is therefore trumped by eligibility.

The explication of eligibility in terms of joint carving suggests the following rough distinction between substantive and verbal disputes: In a substantive dispute, we will find a candidate meaning that carves nature better than its competitors while a verbal disputes involves a plurality of equally eligible candidate meanings. Even if we accept this characterization, we clearly do not have an answer to the demarcation problem. Instead, the demarcation problem reappears in form of the question how we can figure out whether there is a plurality of equally eligible candidate meanings.

One possibility is to explicate joint carving in terms of similarity. A candidate meaning carves nature at its joints if it marks objective similarities in nature. While there is an infinite number of possible candidate meanings, only a limited number of candidate meanings mark objective similarities in nature. Unfortunately, this appeal similarity will be of little help in the discussion of scientific ontologies. In the discussion of natural kinds (4.4), I have argued that there is not one fundamental account of scientific kinds such as species, cognitive processes, or intelligence because the importance of shared properties depends on explanatory interests of scientists.

This lesson extends to judgments about similarity. Whether the members of the biological or ecological species concept are overall more similar depends on whether we consider "being able to produce fertile offspring" or "sharing the same ecological niche" to be more relevant for overall similarity. And whether we consider "being able to produce fertile offspring" or "sharing the same ecological niche" to be more relevant for overall similarity depends on what we are *interested* in and what we want to *do* with the concepts.

Even if we accept similarity as a guide to joint carving, it therefore seems that the life sciences usually come with a plurality of equally eligible ways of carving nature. We therefore end up with a plurality of equally joint-carving scientific ontologies in cases such as species, cognition, or intelligence. As I pointed out in the discussion of natural kinds, it is important not to misunderstand this claim. To say that overall similarity *also* depends on our interests is not to say that it *only* depends on our interest. This point is often ignored, for example by Sider, who claims that the invocation of our interests makes similarity "merely a reflection of something about us" (Sider 2012, 18). But there is an important difference in similarity *merely* being about us and similarity *also* being about us. Many philosophical debates about similarity revolve around two extreme and implausible positions. One extreme is an anti-realism or relativism that claims that similarity is *merely* about us. The other extreme is a strong ontological realism that similarity is *merely* about reality.

Radical anti-realists and relativists miss that similarities are not only about us. Two tigers are similar in countless (anatomical, behavioral, genetic, phylogenetic…) interrelated aspects, no matter whether or not we are interested in them. At the same time, critics of conceptual relativity are wrong in assuming that these similarities can be mapped onto exactly one interest-independent notion of overall similarity that would allow us to identify exactly one interest-independent fundamental biological ontology. Whether the members of an ecological or biological species are overall more similar depends on whether we consider "being able to produce fertile offspring" or "sharing the same ecological niche" to be more relevant. And we cannot evaluate the relevance of properties without taking our interests into account. We need *both*: shared biological features that are independent of our interests, *and* a notion of relevance that presupposes the existence of epistemic and/or non-epistemic interests.

This account of scientific kinds suggests that similarity-judgments will not lead to a demarcation criterion that undermines conceptual relativity but rather to a liberal ontological pluralism in scientific practice. Of course, many metaphysicians will reject this diagnosis. For example, Sider would probably disagree with my claim that conceptual relativity is ubiquitous in science as he suggests that multiple equally candidate meanings are a "relatively uncommon occurrence" (2012, 48). Furthermore, critics of conceptual relativity may also accept conceptual relativity in science but still insist on the ideal of one fundamental philosophical ontology. One way or another, my discussion suggests that the demarcation problem does not only affect proponents of conceptual relativity but is at least as challenging for its critics.

5.5   From Conceptual Relativity to a Pluralist Theory of the Mind?

Let us take stock. The goal of the previous chapters has been to defend conceptual relativity and its pluralist implications. Conceptual relativity is a controversial position in philosophy as it rejects the ideal of exactly one fundamental ontology. After presenting some common philosophical arguments for conceptual relativity, I have suggested that conceptual relativity is ubiquitous in the empirical sciences as exemplified in debates about entities such as species, cognitive processes, or intelligence. The ubiquity of conceptual relativity leads to both a positive and a negative challenge of the ideal of one fundamental ontology: on the one hand, the examples from the empirical sciences suggest that conceptual relativity can be an unproblematic aspect of successful science and is entirely compatible with a moderate realism. The rejection of the ideal of one fundamental ontology therefore does not lead to an unacceptable anti-realism or relativism. On the other hand, case studies from the empirical sciences also challenge the ideal of exactly one fundamental ontology by raising the question why we should stick to this ideal in philosophy if a plurality of equally correct ontologies is part of the everyday business of successful science.

Critics of conceptual relativity can reply to this challenge by arguing that conceptual relativity has unacceptable implications. For example, one can argue that conceptual relativity leads to the absurd and self-defeating claim that *all* existence disputes are verbal disputes that can be resolved through the distinction between conceptual frameworks. I have argued that this objection implies a "demarcation problem" that has to be addressed by both proponents and critics of conceptual relativity. Furthermore, I have suggested that the most plausible answers to the demarcation problem actually support conceptual relativity and a rather liberal ontological pluralism in the empirical sciences.

Even if we accept all of this for the sake of the argument, one may wonder why any of my claims should be relevant for a pluralist theory of mind. There are two largely distinct answers to this question. On the one hand, my discussion of conceptual relativity in psychology and cognitive science suggests a straightforward answer: there is not one fundamental ontology of the mind but a plurality of equally fundamental psychological and cognitive ontologies. Instead of wondering whether entities such as a general intelligence or extended cognitive processes really exist, we should endorse the plurality of ontologies that we find in scientific practice. This position leads to deflationist interpretations of many philosophical debates in cognitive science and it also provides a more positive account of ontological issues in the cognitive sciences. For example, psychiatry is a discipline that involves a variety of ontologies and many ontological disputes such as the heated controversies about the most recent revision of the *Diagnostic and Statistical Manual of Mental Disorders* (DSM-5) (e.g. Casey et al. 2013; Frances 2013; Nemeroff et al. 2013). A pluralist approach suggests that we should not worry about the question what mental disorders *really* exist but engage in debates about the epistemic and non-epistemic virtues of different psychiatric ontologies (Zachar and Kendler 2007; Sisti et al. 2013, cf. Lemeire 2014).

On the other hand, my preliminary discussion of a pluralist theory of the mind in the introduction clearly promised more by also challenging debates about reduction and explanatory gaps in philosophy of mind. For example, I have proposed a pluralist approach that challenges explanatory gap problems in philosophy of mind by treating the scope of ontological and epistemic unification an open empirical question. However, it is at least not immediately clear that my discussion of conceptual relativity implies such an ambitious pluralism. For example, one could accept a plurality of psychological and cognitive ontologies but still insist that psychological and cognitive entities have to be explained in terms of more fundamental biological or physical entities. In the following chapters, I will present two arguments that connect conceptual relativity with debates about reduction and that lead to a more ambitious pluralism in philosophy of mind.

References

Carroll, Lewis. 1951. *Lewis Carroll's Alice in Wonderland: And Other Favorites*. New York: Washington Square Press.

Casey, B. J., Craddock, N., Cuthbert, B. N., Hyman, S. E., Lee, F. S., & Ressler, K. J. 2013. DSM-5 and RDoC: Progress in Psychiatry Research? *Nature Reviews Neuroscience* 14 (11): 810–814.

Chalmers, David. 2011. Verbal Disputes. *Philosophical Review* 120 (4): 515–66.

Davidson, Donald. 1984. *Truth and Interpretation*. Oxford: Calderon Press.

Hirsch, Eli. 2005. Physical Object Ontology, Verbal Disputes, and Common Sense. *Philosophy and Phenomenological Research 70* (1): 67–97.

Hirsch, Eli. 2008. Ontological Arguments: Interpretive Charity and Quantifier Variance. In *Contemporary debates in metaphysics,* ed. Ted Sider, John Hawthorne and Dean W Zimmerman, 367–381. Malden, MA: Blackwell Publishing.

Hirsch, Eli. 2011. *Quantifier Variance and Realism: Essays in Metaontology*. Oxford: Oxford University Press.

Hirsch, Eli. 2013. Charity to Charity. *Philosophy and Phenomenological Research* 86 (2): 435–442.

Frances, Allen. 2013. *Saving Normal: An Insider's Revolt Against Out-of-Control Psychiatric Diagnosis, DSM-5, Big Pharma, and the Medicalization of Ordinary Life*. New York: William Morrow.

Jackson, Brendan Balcerak. 2013. Metaphysics, Verbal Disputes and the Limits of Charity. *Philosophy and Phenomenological Research* 8 (2): 412–434.

Jenkins, Carrie. 2014. Merely verbal disputes. *Erkenntnis 79* (1): 11–30.

Lemeire, Olivier. 2014. Soortgelijke stoornissen. Over nut en validiteit van classificatie in de psychiatrie. *Tijdschrift voor Filosofie* 76 (2): 217–246.

Manley, David. 2009. Introduction: A Guided Tour of Metametaphysics. In *Metametaphysics: New Essays on the Foundations of Ontology,* eds. David John Chalmers, David Manley and Ryan Wasserman, 1–47 Oxford: Oxford University Press.

Melville, Herman. 2013. *Moby Dick: Or, The Whale*. New York: Modern Library.

Nemeroff, C. B., Weinberger, D., Rutter, M., MacMillan, H. L., Bryant, R. A., Wessely, S., & Lysaker, P. 2013. DSM-5: a Collection of Psychiatrist Views on the Changes, Controversies, and Future Directions. *BMC medicine* 11 (1): 202.

Sider, Ted. 2001. Criteria of personal identity and the limits of conceptual analysis. *Noûs* 35 (15): 189–209.

Sider, Ted. 2009. Ontological Realism. In *Metametaphysics: New Essays on the Foundations of Ontology,* eds. David John Chalmers, David Manley and Ryan Wasserman, 384–423. Oxford: Oxford University Press.

Sider, Ted. 2012. *Writing the Book of the World*. Oxford: Oxford University Press.

Sisti, Dominic, Michael Young, and Arthur Caplan. 2013. Defining mental illnesses: can values and objectivity get along? *BMC psychiatry* 13 (1): 1–4.

Warren, Jared. 2014. Quantifier Variance and the Collapse Argument. *The Philosophical Quarterly* 65 (259): 241–253.

Wilson, Neil L. 1959. Substances Without Substrata. *The Review of Metaphysics* 12 (4): 521–39.

Zachar, Peter, and Kenneth Kendler. 2007. Psychiatric disorders: a conceptual taxonomy. *American Journal of Psychiatry* 164 (4): 557–565.

Part III
From Conceptual Relativity
to Vertical Pluralism

Chapter 6
The Argument from Horizontal Pluralism

Recall Putnam's example of a room with "a chair, a table on which there are a lamp and a notebook and a ballpoint pen, and nothing else" (1988, 110). In the last chapter, I used this example as an illustration of conceptual relativity in ordinary language. Ordinary language provides a plurality of equally legitimate ways of counting objects and therefore a plurality of equally legitimate ordinary ontologies. For example, we can count the book as one object but we can also count its pages as different objects. While all of this is well-known from the last chapter, Putnam assumes that there is another lesson to learn from the example: we can describe the room not only in ordinary language, we can also describe it in terms of scientific languages such as particle physics. Furthermore, Putnam argues that the diagnosis of conceptual relativity should be extended to these cases. In discussing the relation between ordinary and microphysical descriptions, we should also avoid the assumption of one absolute description of reality that implies the ideal of one fundamental ontology. Instead, we should accept a plurality of conceptual resources that comes with a plurality of ontologies that do not need to be unified through a reduction to exactly one fundamental (e.g. microphysical) ontology.

Putnam's claims about the relation between ordinary and microphysical descriptions suggest that we have to distinguish two different kinds of conceptual relativity. On the one hand, there are examples of conceptual relativity in one (e.g. biological or psychological) domain of inquiry. My case studies of species, extended cognition, and intelligence illustrate this kind of conceptual relativity in the empirical sciences. On the other hand, Putnam suggests that there are also examples of conceptual relativity across *different* domains, such as ordinary language and particle physics or psychology and biology.[1] In *Ethics without Ontology*, Putnam acknowledges the need to distinguish between those two types of cases:

---

[1] While it is often convenient to make this distinction between different domains or levels by pointing to different scientific disciplines, talk about "domains" or "levels" should be taken with a grain of salt. Many scientific research projects have a multilevel character in the sense that they involve multiple levels of organization. Craver (2007) discusses the multilevel character of scientific

© Springer International Publishing Switzerland 2015
D. Ludwig, *A Pluralist Theory of the Mind*, European Studies
in Philosophy of Science 2, DOI 10.1007/978-3-319-22738-2_6

> In *Representation and Reality* I counted the fact that we might describe "the contents" of a
> room very differently by using first the vocabulary of fundamental physical theory and then
> again the vocabulary of tables and lamps and so on as a further instance of conceptual rela-
> tivity and this, I now think, was a mistake, although it is an instance of a related and wider
> phenomenon I should have called *conceptual pluralism*. The fact that the contents of a room
> may be partly described in two very different vocabularies cannot be an instance of concep-
> tual relativity in the sense just explained, because conceptual relativity always involves
> descriptions which are cognitively equivalent […] but which are incompatible if taken at
> face value. (Putnam 2004, 48)

Putnam's distinction provides a helpful opportunity to clarify the terminology of
debates about pluralism. In the introduction, I argued that conceptual pluralism
should be distinguished from a merely epistemological pluralism that has no ontological consequences and a strong metaphysical pluralism that argues for exactly
one fundamental pluralist ontology such as Popper's theory of three worlds. In the
last chapter, I argued that conceptual relativity implies conceptual pluralism. For
example, conceptual relativity in the species debate implies a plurality of equally
fundamental species ontologies that reflect different conceptual choices in biology.
This use of "conceptual pluralism" differs from the Putnam quote that distinguishes
between conceptual relativity and conceptual pluralism. Despite this terminological
difference, Putnam clearly makes a valid distinction that will be of crucial importance for the discussion in this chapter. Furthermore, his distinction can be expressed
through the more established terminology of "horizontal" and "vertical" pluralism.[2]
Horizontal pluralism is the claim that there can be different but equally fundamental
descriptions in one domain, while vertical pluralism assumes different but equally
fundamental descriptions across domains. Horizontal pluralism corresponds with
Putnam's more recent and restricted use of "conceptual relativity" while vertical
pluralism corresponds with his use of "conceptual pluralism". Figure 6.1 offers a
simple illustration of the distinction.

Horizontal pluralism claims that there can be different but equally fundamental
descriptions in terms of ordinary language (O1 & O2) or particle physics (P1 & P2),
and describes the relation represented by the dashed lines. Vertical pluralism claims
that the same situation can be described on different but equally correct conceptual

---

explanations in detail. For example, an explanation of fluid homeostasis involves entities such as
"behaviors of organisms (drinking), drives (thirst), the working of bodily organs (conservation of
urine in the kidneys), the flux of bodily molecules (such as the pituitary's release of vasopressin),
and swarms of ions (the concentration of salt in the blood)" (9). Bapteste and Dupré (2013) provide
another example by arguing that microbiological research is increasingly concerned with dynami-
cal systems that involve "causal interactions between entities from different levels of biological
organization" (379).

[2] As far as I know, the distinction between horizontal and vertical forms of pluralism has been
introduced by Price (1992). For more recent accounts, see Lynch (2001, 6–8) and Mitchell (2003).
Furthermore, the horizontal-vertical distinction is clearly related to interlevel-intralevel distinction
in debates about theory reductions (e.g. Wimsatt and William 1976).

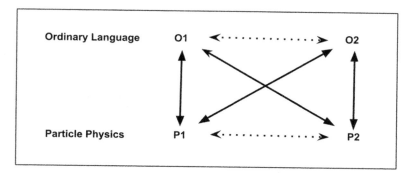

Fig. 6.1 Schematic illustration of the relation between horizontal and vertical pluralism

levels (P1 & O1, P1 & O2, P2 & O1, P2 & O2) and describes the relation represented through the solid lines.[3]

If we use the distinction between horizontal and vertical pluralism, the examples
from the last chapter all turn out to be cases of horizontal pluralism. Horizontal
pluralism is philosophically interesting because ontologies on a horizontal scale
often seem to contradict each other: how can there be exactly five objects and
exactly seven objects in the same room at the same time? The puzzlement vanishes
when we realize that there are different correct ways to count objects. In this sense,
horizontal pluralism is of considerable deflationary power, as it can dissolve disputes in ordinary language, science, and philosophy by differentiating between different conceptual frameworks.

While horizontal pluralism is an interesting and controversial thesis in metaphysics and philosophy of science, a pluralist theory of the mind obviously requires
vertical pluralism. In the following chapters, I present two arguments for vertical
pluralism. According to the *argument from horizontal pluralism* (this chapter), the
plurality of equally fundamental descriptions on the horizontal level implies plurality on a vertical level. According to the *argument from ontological non-
fundamentalism* (Chap. 7), a global reductivism is tied to the ideal of exactly one
fundamental ontology. If we do not presuppose that everything must be explained in
terms of one fundamental ontology, we should adopt a relaxed non-reductivism
(opposed to both reductivism and anti-reductivism) that considers the scope of
reductive explanations an open empirical question.

---

[3] While the distinction between horizontal and vertical pluralism is often heuristically helpful, it is
not without problems. Most importantly, one may worry that the idea of a vertical scale comes with
a problematic layer picture of a mereological hierarchy of clearly distinguished ontological levels.
However, I do not use the horizontal-vertical distinction to postulate such a problematic metaphysical picture but to refer to the rather basic observation that many of our ontologies involve
aspects of composition and scale-relativity (cf. Eronen's "deflationary account" of levels 2013).
Furthermore, a breakdown of the horizontal-vertical distinction would arguably only strengthen
my argument as I would not have to make a separate case for vertical pluralism.

6.1   Does Horizontal Pluralism Imply Vertical Pluralism?

It would be very convenient if we could show that horizontal pluralism (or "conceptual relativity" in the narrow sense of Putnam's more recent terminology) implies vertical pluralism. Given the presentation of horizontal pluralism from the last chapter, vertical pluralism and maybe even a pluralist theory of the mind would come as a free lunch. And indeed, Putnam claims that "while conceptual relativity implies [vertical] pluralism, the reverse is not the case" (2004, 48). Unfortunately, he does not offer a clear argument for his claim that horizontal pluralism implies vertical pluralism. Putnam insists that, contrary to horizontal pluralism, vertical pluralism does not involve *prima facie* contradictions. Consider the following examples.

(12) There are exactly three objects in Putnam's universe with three individuals
(13) There are exactly seven objects in Putnam's universe with three individuals
(32) There is a chair in the room
(33) There are elementary particles in the room

A horizontal pluralist who argues that (12) and (13) are equally correct has to explain how they can be true even if they seem to contradict each other. Conceptual relativity offers an explanation because it considers sentences such as (12) and (13) to be true relative to the choice of an ontology. The comparison of (32) and (33) does not involve a contradiction and therefore does not raise the question of how it is possible that both (32) and (33) are true. Given this difference between (12) vs. (13) and (32) vs. (33), horizontal pluralism seems to be the more ambitious claim as horizontal pluralists have to deal with *prima facie* contradictions.

However, even if horizontal pluralism is more ambitious in the sense that it involves *prima facie* contradictions, we do not have a clear argument for the claim that horizontal pluralism *implies* vertical pluralism. Horizontal pluralism might be more ambitious in the sense just explained but there might be other reasons to accept horizontal pluralism and to reject vertical pluralism.

Furthermore, it is not difficult to imagine a philosopher who accepts horizontal pluralism and rejects vertical pluralism. Consider a reductive physicalist who accepts a traditional picture of microreduction: psychology can be reduced to biology, biology can be reduced to chemistry, and chemistry can be reduced to microphysics. Under the assumption of this reductive physicalism, vertical pluralism is wrong because only a physical ontology is truly fundamental. However, it is not immediately clear why a reductive physicalist has to reject horizontal pluralism. Why should she not accept that there are different and equally fundamental ontologies on a horizontal scale, but insist that they are still reducible to a fundamental physical ontology? For example, a reductive physicalist might accept that there are different and equally fundamental accounts of species or cognition but still insist that every biological and cognitive entity can be explained in terms of fundamental physical entities. It is certainly not obvious that this is an inconsistent position and it seems that any convincing case for vertical pluralism has to present an argument against global reductionism.

In his book *Truth in Context*, Michael Lynch (2001) acknowledges that a vertical pluralist has to reject reductionism, but still insists that horizontal pluralism implies vertical pluralism. Lynch distinguishes between local and global versions of pluralism. Local pluralism is restricted to a particular type of discourse, while global pluralism is a claim about every kind of discourse. According to Lynch, local horizontal pluralism entails local vertical pluralism, and global horizontal pluralism entails global vertical pluralism. Here is his argument:

> With this distinction in hand, we may now ask whether a local horizontal pluralism entails a local vertical pluralism. Let us take moral facts as the example. Does horizontal moral pluralism entail vertical moral pluralism? The question, in other words, concerns whether moral relativism implies that the (relative) moral facts are irreducible to physical facts. To say that one type of facts is reducible to another type of facts is to imply that the former can be completely explained in terms of the latter. But once moral facts are relativized to cultures or practices, then it seems that no set of physical facts alone will be able to capture or explain what is the case at the moral level. For any explanation of the moral facts would have to appeal to the culture or practice those facts were relative to. Relative moral facts, then, would seem irreducible to underlying physical facts, and hence local horizontal pluralism would appear to imply vertical pluralism (Lynch 2001, 7).

Lynch's example of moral facts is slightly unfortunate for at least two reasons. First, moral facts pose a general challenge to reductionism that is independent from horizontal pluralism: how are moral – or, more generally, normative – facts possible if the fundamental physical facts are not normative? Second, many physicalists react to this challenge with a deflationary account of moral facts. Somehow, moral facts are not real facts and therefore there is no need for reduction.

I think that the discussion will benefit from a consideration of non-normative cases of horizontal pluralism. In the last chapter, I argued that conceptual relativity extends to psychology and that there are different but equally fundamental accounts of intelligence (Sect. 4.3). We can decide to use an ontology that only accepts a general intelligence or we can decide to use an ontology of multiple intelligences. Furthermore, it depends on our explanatory interests which cognitive abilities we include in discussions about intelligence. For example, Howard Gardner's concept of multiple intelligences includes a musical intelligence, while traditional psychometric accounts do not measure musical abilities. In this sense, psychological entities such as intelligences and psychological facts such as "Paul has an IQ of 97" are relative to our ontological choices. Lynch's presentation suggests that this local horizontal pluralism in psychology implies the irreducibility of psychology. Here is one possible reconstruction of the argument:

1. Psychological facts are relative to the ontological choices of psychologists.
2. If psychological facts are relative to the ontological choices of psychologists, they are not implied by biology or physics.
3. Reductionism in psychology requires that psychological facts are implied by biology or physics.
∴. Reductionism in psychology is false.

Of course, we can also generalize this argument and defend global vertical pluralism:

1. Non-physical facts are relative to our ontological choices.
2. If non-physical facts are relative to our ontological choices, then they are not implied physics.
3. Reductionism requires that all facts are implied by physics.
∴ Reductionism is false.

How should a reductionist react to these arguments? One possibility is to reject the first premise of the argument, which would be tantamount to the rejection of horizontal pluralism. If a reductionist wants to show that her position is compatible with horizontal pluralism, she has to either deny the second or the third premise.

Unfortunately, I think that there is a rather obvious problem with the third premise of the argument. Even classical reductionist models do not claim that "all facts are implied by physics", but acknowledge the need of bridge principles that connect physical and non-physical levels. They do not claim that all facts are implied by physics but by physics *in conjunction with appropriate bridge principles*. A classical example of the function of bridge principles in reductive explanations is the case of water and H_2O.[4] The reduction of water requires the reductive explanation of facts such as:

(34) Water freezes at 0 °C
(35) Water boils at 100 °C

If we want to reduce water to H_2O, we need to infer macroscopic facts about water such as (34) and (35) from chemical facts. However, concepts such as "freezing" or "boiling" are not part of the chemical vocabulary, which makes it hard to see how (34) and (35) could be inferred from any chemical description. Consider a chemical description of the impact of different temperatures on H_2O molecules. At 0 °C, the forces between H_2O molecules become so strong that they form densely packed crystalline structures. As a consequence, the molecules cannot move freely and a comparably large force is necessary to break them apart. It seems obvious that chemical facts like these offer a sufficient explanation of (34). However, a chemical description cannot entail (34) because "freezing" is not part of the chemical vocabulary. In order derive (34) from a chemical description we need so called "bridge principles," which connect the chemical vocabulary with ordinary concepts. If we have a bridge principle according to which x freezes if certain chemical conditions are met, and we have a chemical description of these conditions, then we can derive (34) from the chemical description in conjunction with the bridge principle.

The necessity of bridge principles suggests a simple objection against my interpretation of Lynch's anti-reductionist argument. Indeed, psychological facts are relative to the ontological choices of psychologists and they are not implied

---

[4] The *locus classicus* for philosophical debates about bridge principles is Nagel (1979, Chap. 11). For an excellent discussion of the allegedly obvious reduction water to H_2O, see Chang (2012).

by biology. However, they are implied by biology *in conjunction with appropriate bridge principles*. For example, psychological facts about intelligence are relative to the choice of an account of intelligence but still reducible to biological facts under the assumption of appropriate bridge principles. In the same way as we need bridge principles that connect ordinary concepts such as "freezing" or "boiling" with chemical concepts, we need bridge principles that connect psychological concepts such as "IQ", "general intelligence", "verbal intelligence", "musical intelligence" with neuroscientific concepts. A plurality of accounts of intelligence therefore only illustrates the need for a plurality of bridge principles but does not threaten reductionism. For example, we can choose to include or exclude musical abilities in our accounts of intelligence and in this sense we can choose between different cognitive ontologies. However, these ontological choices do not threaten reductionism as we can reduce facts about intelligence no matter whether we choose a more liberal or a more restricted account of intelligence. Different accounts of intelligence simply require different neuroscientific explanations and bridge principles.

6.2 Dupré's Promiscuous Realism

Both Putnam and Lynch claim that horizontal pluralism implies vertical pluralism. Putnam points out that, contrary to horizontal pluralism, vertical pluralism does not involve *prima facie* contradictions. However, it is far from clear how this observation could be turned into an argument for the claim that horizontal pluralism *implies* vertical pluralism. Lynch offers an argument according to which conceptual relativity on the horizontal scale implies that facts are not reducible on a vertical scale. At least my initial reconstruction of Lynch's argument remained unsatisfying as it did not consider the role of bridge principles in vertical reduction. At the same time, a closer look at this argument indicates how a successful argument for vertical pluralism might look. It would be necessary to show that horizontal pluralism is incompatible with reductionism *even if* we take bridge principles into account.

In order to improve the argument with a discussion of bridge principles, it is helpful to have a look at John Dupré's "promiscuous realism," which combines a resolute pluralism on the horizontal and vertical scale. A discussion of Dupré's promiscuous realism requires some terminological clarifications as Dupré does not use the labels "horizontal pluralism" or "conceptual relativity," but presents "essentialism" as one of his main targets. According to Dupré, "essentialism" is the idea that natural kinds can be understood in terms of essences: every member of a natural kind has the same essence. Dupré presents essentialism in the context of biological case studies such as the species concept, where essentialists assume that every member of a species must share a common essence. Although essentialism is widely rejected in contemporary biology, one might still be tempted to assume that a microstructural (e.g. genetic) structure can serve as the essence of a species. Dupré points out that essentialism is tempting but does not match biological reality.[5] In the case of genetics, intraspecific variation comes with genetic variation and there is little hope in finding invariant cores that separate species from one another.

Dupré's rejection of biological essentialism invites the obvious objection that we can always *define* essences. Recall the different species concepts from the last chapter. One may argue that these species concepts trivially define essences. According to the biological species concept, members of a species share the essential property of being able to interbreed. According to the ecological species concept, members of a species share the essential property of inhabiting the same niche. According to the phylogenetic species concept, members of a species have the essential property of sharing the same lineage. Does this not make the existence of essences a triviality?

In order to answer this question, we need to know more about how Dupré understands "essence." Dupré clarifies that his anti-essentialism is not directed against a trivial concept of essence but against what is traditionally called a "real essence." The existence of essences is trivial if we define "essence" as whatever a scientist considers crucial to determine membership to a kind. However, such a notion would imply that essences are relative to the theories and interests of scientists. Contrary to these theory- and interest-relative essences, real essences are supposed to be discovered in nature and to be independent of our interests. As Dupré puts it: "The existence of such real essences would imply that there is some unique, privileged scheme of classification, which assigns everything to a class defined by common possession of the appropriate essence. While the existence of such a privileged scheme might be compatible with the existence of disparate categories for the rough-and-ready purposes of everyday life, it surely does not entail that only the one privileged scheme is adequate for the purposes of science" (1993, 60).

In the case of species, this kind of essentialism implies exactly one correct way of determining species membership. However, Dupré points out there is no such real essence but a variety of possible ways to shape biological kinds. "There is no God-given, unique way to classify the innumerable and diverse products of the evolutionary process. There are many plausible and defensible ways of doing so" (1993, 57).

The similarities between Dupré's presentation of anti-essentialism and my presentation of conceptual relativity in biology are striking and his anti-essentialism would not greatly advance the present discussion, if he would limit pluralism to the horizontal scale. However, Dupré clarifies in the second chapter of his book that promiscuity is necessary on both the horizontal and the vertical scale: "Whereas the first part of this book was intended to establish the existence of many overlapping but equally real classifications of objects on one level of organization (the biological), here I want to argue for an equally liberal pluralism across structural levels of organization" (1993, 89). Furthermore, Dupré insists that pluralism on the vertical

---

[5] While the non-existence of traditional essences has become a truism in philosophy of biology, some philosophers have adopted the label "essentialism" to describe their positions (cf. Ereshefsky 2010 for a helpful overview). Most contemporary essentialisms (e.g. Okasha 2002 and LaPorte 2004) postulate relational properties as essences and are therefore clearly compatible with Dupré's rejection of "real essences". Devitt's (2008) account differs by requiring intrinsic essences.

scale is a direct consequence of the rejection of essentialism. "In an important sense this observation [that many scientific kinds lack essential properties] is sufficient for the refutation of reductionism" (1993, 105).

How does Dupré's argument run? As we have seen so far, the rejection of essentialism implies pluralism on a horizontal scale. Biological kinds are not grounded in real essences, but reflect the explanatory interests of biologists. In this sense, a scientific "system of classification is typically an inextricable part of the science to which it applies" (1993, 103). But how does this horizontal plurality transfer to plurality on the vertical scale? The crucial premise in Dupré's argument for vertical pluralism is that the lack of real essences undermines the hope to find coextensive kinds on a vertical scale. Given the assumption that biological kinds have essences, we could hope to identify the essences with specific chemical kinds and finally with physical kinds. However, if there are many different legitimate ways to shape biological kinds relative to the explanatory interests of biologists, then there is no reason to believe that biological kinds will correspond to coextensive chemical or even physical kinds. Instead, many legitimate biological kinds will have little to do with physical structure and correspond to vastly different physical kinds. And if biological kinds do not correspond to coextensive physical kinds, we cannot reduce biological kinds and biological facts to a physical level.

The argument becomes clearer when we take Dupré's discussion of bridge principles into account. According to Dupré, reduction requires "bridge principles (or bridge laws) identifying the kinds of objects at the reduced level with particular structures of the object at the reducing level" (1993, 88). Let us consider a simple example such an object d that belongs to the biological kind *Taraxacum officinale* (common dandelion). Given that d has the physical structure P, we can attempt to formulate the following bridge principle:

(B1) if x has the physical structure P, x is a common dandelion

Our knowledge that d has in fact the physical structure P in conjunction with the bridge principle (B1) implies that d is a common dandelion. What, then, is the problem with reductionism? According to Dupré, the problem is that (B1) is not a legitimate bridge principle, as bridge principles have to be biconditionals (1993, 105) and would have to have the following form:

(B2) x is a common dandelion, if *and only if it* has to the physical structure P

If essentialism were true, the availability of bridge principles like (B2) would be a realistic possibility. If there were a real essence of *Taraxacum officinale*, we could hope for that essence to be identical with P so that every common dandelion would have the same essential physical structure P. And if every common dandelion would have the same essential physical structure P, we could formulate a bridge principle with necessary and sufficient conditions and reduce the biological kind *Taraxacum officinale*.

However, Dupré argues that essentialism is wrong and that there are many equally legitimate ways of dividing the biological realm into scientific kinds. A system of classification is an "inextricable part of the science to which it applies" and, therefore, members of the same non-physical kind will often belong to different physical kinds. Consider two common dandelions d1 and d2, where d1 has the physical structure P1 and d2 has the physical structure P2. According to Dupré, P1 and P2 may very well have no interesting physical properties in common that distinguish P1 and P2 from the physical structures of other plants of the genus *Taraxacum*. Therefore, we cannot find a biconditional bridge principle that connects *Taraxacum officinale* to a physical kind, and many biological kinds will remain irreducible. In other words, the rejection of essentialism and the endorsement of horizontal pluralism ensures vertical autonomy as kinds are shaped by the unique logic of their scientific domain and will not always correspond to coextensive physical kinds.

In the last section, I criticized Lynch's argument by suggesting that physical descriptions *in conjunction with bridge principles* could be sufficient for reductions even under the assumption of horizontal pluralism. Dupré's promiscuous realism illustrates how a vertical pluralist can argue that appropriate bridge principles will not be available. Bridge principles have to be biconditionals but horizontal pluralism undermines the hope that we will be able find biconditional bridge principles that connect physical and non-physical kinds.

6.3   Bridge Principles and Notions of Reduction

The argument I have presented so far connects horizontal and vertical pluralism by arguing that pluralism on the horizontal scale implies that we won't always find coextensive kinds on a vertical scale.[6] For example, there are no biconditional bridge principles that connect species with physical kinds. Although *Taraxacum officinale* is an interesting biological kind, it does not correspond to an interesting physical kind. The same consideration arguably applies to psychology. Even if we restrict ourselves to human psychology, it remains highly unlikely that every legitimate psychological kind corresponds to a coextensive neural kind (cf. Sect. 6.4).

The argument from horizontal pluralism states that we will not always find coextensive kinds on a vertical scale. However, is this really enough to support vertical pluralism? Perhaps the requirement of coextensive kinds was misguided from the very beginning and reductionists can react to the argument by lowering their demands. An obvious strategy would be to formulate one-way bridge principles that only require sufficient but not necessary conditions (Richardson 1979). So far, we have assumed that reductionism requires that physical (P) and non-physical kinds (N) are connected through biconditional bridge principles:

$$P \leftrightarrow N$$

---

[6] Of course, this argument could also be made with biological individuals or properties instead of kinds. For example, see Clarke (2013) for a helpful account of the multiple realizability of organisms.

But why is it not enough to formulate a large number of one-way bridge principles that connect physical with non-physical kinds?

$$P_1 \rightarrow N$$
$$P_2 \rightarrow N$$
$$P_3 \rightarrow N$$
....
$$P_n \rightarrow N$$

These one-way bridge principles would formulate sufficient conditions and make N derivable from the physical description in conjunction with bridge principles. And if N remains derivable, one-way bridge principles may appear to be all we need to defend a reasonable reductionism. Furthermore, the availability of one-way bridge principles would even make the construction of a disjunctive biconditional bridge principle possible:

$$P_1 \vee P_2 \vee P_3 \ldots P_n \leftrightarrow N$$

In other words: if a non-physical kind does not correspond to *one* physical kind, why can we not simply reduce it to *several* physical kinds? For example, if a species does not correspond to a physical kind P, it will still correspond to a large number of different physical kinds $P_1, P_2, P_3 \ldots P_n$. And why can we not reduce the species to the heterogeneous physical kind $(P_1 \vee P_2 \vee P_3 \ldots P_n)$?

One possible answer to this question is that the "reduction" to a heterogeneous disjunctive physical kind would actually vindicate vertical pluralism by acknowledging the disunity of different domains. Of course, there is little point in a verbal discussion about the correct meaning of "reduction" and we can define "reduction" in a way that biconditional bridge with disjunctive physical kinds are sufficient for reduction. Given this definition of reduction, however, the interesting patterns of the reduced level will often not be detectable on the reducing level. Consider a few uncontroversial biological facts about the common dandelion *Taraxacum officinale*:

(35) All members of *Taraxacum officinale* are asexual species
(36) The offspring of *Taraxacum officinale* is usually genetically identical to the parent plant
(37) *Taraxacum officinale* is a ruderal species that quickly colonizes disturbed lands

If biological kinds such as *Taraxacum officinale*, "asexual organism," or "ruderal species" would be coextensive with physical kinds, we could reduce (35)–(37) to a physical level. For example, if *Taraxacum officinale* were coextensive with the physical kind F and "asexual organism" were coextensive with the physical kind G, we could reduce (35) to the claim that all members of F are also members of G.

If there are no coextensive physical kinds, however, then we cannot preserve interesting biological facts such as (35)–(37) on a physical level. It is obvious that the disjunctive move will not change anything about this situation. Let us assume for the sake of the argument that we can identify *Taraxacum officinale* with the disjunctive physical kind $(F_1 \vee F_2 \vee F_3 \ldots F_n)$ and "asexual organism" with the disjunctive physical kind $(G_1 \vee G_2 \vee G_3 \ldots G_n)$. This would allow us to reformulate

(35) as "Every member of the disjunctive kind $(F_1 \vee F_2 \vee F_3 \ldots F_n)$ is also a member of the disjunctive kind $(G_1 \vee G_2 \vee G_3 \ldots G_n)$." But this reformulation only reinforces the idea of vertical pluralism. (35) is an interesting biological fact and biological kinds, such as *Taraxacum officinale* or "asexual organism," are natural kinds at least in the sense that they are of theoretical importance and explanatory power in biology. As members of the disjunctive physical kinds do not have any unique and interesting physical properties in common,[7] interesting biological facts such as "All members of *Taraxacum officinale* are asexual species" would be lost in a reduction to physical facts that involve disjunctive physical kinds.

It may helpful to recall the relevance of the present discussion for the overall aim of this book. My goal is not to propose a general account of reduction and my naturalist methodology may also lead to the conclusion that different accounts of reduction may turn out to be useful in different areas of scientific practice. Instead, my goal is to address the question whether there is any reason to endorse a reductionism *that is strong enough* to undermine vertical pluralism by aiming at global ontological and epistemic unification. This goal is entirely compatible with more moderate notions of "reduction" that may be useful in one way or another but do not challenge vertical pluralism.

To reinforce this point, consider recent efforts to rehabilitate theory reductions in the tradition of Nagel (1979) and Schaffner (e.g. 1967, 1976). Dizadji-Bahmani, Frigg, and Hartmann (2010) have argued that the Nagelian theory of reduction has received a lot of bad press for bad reasons and that the core elements of the Nagel-Schaffner model stand uncorrected. They defend these claims not only in the context of the reduction of thermodynamics to statistical mechanics but also address arguments from multiple realization according to which properties in a reducing theory T_f do not correspond to coextensive properties of a reduced theory T_p.

The non-availability of coextensive properties (or kinds) usually leads to ontological and epistemological worries regarding theory reductions. On the ontological side, it is unclear how T_p-properties can be "nothing over and above" T_f-properties. On the epistemological side, it is unclear how heterogeneous T_f-properties could explain T_p-properties. The main strategy of Dizadji-Bahmani, Frigg, and Hartmann is to declare these worries irrelevant for reduction in scientific practice. With regard to ontological worries they acknowledge that it is commonly held that reductions have to show that "T_P-properties are nothing over and above T_f-properties. We

---

[7] Ladyman and Ross (2007) correctly point out that it is often problematic to claim that the types or kinds of the "special sciences" have *nothing* in common with physical types and kinds: "Physical descriptions of the tokens of at least many special science types often have a great deal (and certainly far more than 'nothing') in common. Questions of physical similarity aren't irrelevant to, for example, whether two animals are both vertebrates, or whether two different samples of sediment are clays or oozes" (2007, 50). My formulation of "unique and interesting" differences accommodates this observation: for example, there may be similarities that distinguish members of a vertebrate species from members of invertebrate species but that does not mean that there are *unique* similarities that distinguish them from members of other vertebrate species. Furthermore, even if we would find physical similarities that distinguish members of the same species from members of all other species, they would probably not be *interesting* in the sense that they would be of any importance in the explanation of the unique biological features of species. At best, they would be interesting in the sense that they provide contingent physical "markers" for biological kinds.

believe this to be mistaken. Whether or not the establishment of strict identities is a desideratum for a reduction depends on what one wants a reduction to achieve. If metaphysical parsimony or the defense of physicalism are one's primary goals, then identity may well be essential [but] in science neither of these are very high on the agenda" (2010, 405). With regard to epistemological worries, they declare that "reductions do not ipso facto have to double as explanations. The two core aims of reduction— consistency and confirmation—can be had without adding further items to the list, and reductions are desirable even if they do not serve any other purposes. Explanation, in particular, is nice to have where it can be had, but it is not a sine qua non of reduction" (2010, 407; cf. Walter 2006 for a similar point).

While I am happy to accept this account of reduction,[8] it should be immediately clear that it does not contradict my presentation of vertical pluralism and that it is of little help for philosophers who insist on a reductionism that aims at global onto-logical and epistemic unification. Given the lack of ontological implications (at least with regard to properties), the revived Nagelian account nicely fits my claim that we should consider the scope of ontological unification an open empirical question and avoid the metaphysical presupposition of exactly one fundamental ontology. Given the characterization of explanations as a nice but inessential feature of reductions, the revived Nagelian account is also of little help in supporting claims that explanatory gaps are rare or even non-existent in science. Instead, a successful reduction of T_P does not rule out that there are T_P-properties that are not explicable in terms of T_F-properties.

The discussion of the argument from horizontal pluralism therefore leads to the following picture. Pluralism on the horizontal scale undermines hopes that we will always find coextensive kinds on the vertical scale. While this result challenges ambitious variants of vertical reductionism, one can respond by lowering the demands for successful reduction as suggested by Richardson (1979) as well as Dizadji-Bahmani, Frigg, and Hartmann (2010). Furthermore, it would be easy to add further examples from the current literature on reduction to the discussion. For example, Bickle (2003, 2008) urges us to adopt a "metascientific" account that differs from traditional debates in philosophy of science by letting scientific practice decide what counts as a successful reduction. This metascientific account also motivates Bickle to reject traditional ontological and epistemological criteria for successful reduction and to "let ontological chips fall where they may" (2003, 32). I'm fine with all of this and I have no ambition to engage in a dispute about the correct definition of "reduction". At the same time, none of the mentioned accounts challenge the main point of the argument from horizontal pluralism: horizontal pluralism under-mines reductionism *if* "reductionism" is understood in a philosophically ambitious sense that contradicts pluralism by aiming at ontological and epistemic unification.

---

[8]This does not mean that I expect this account to helpful in *all* areas of scientific practice. For example, it is doubtful that the revived Nagelian model will capture all relevant meanings of "reduction" in science and limits are especially apparent in the life sciences (cf. Kaiser and Marie 2012; Brigandt and Love 2012). One reason is that many reductions in the life sciences are in fact essentially explanatory and therefore much closer to my discussion of reductive explanations in Sects. 6.5 and 6.6.

6.4   Horizontal Pluralism and Multiple Realization

The argument from horizontal pluralism shares important assumptions with well-known arguments from multiple realization. Many traditional cases for multiple realization come from *prima facie* plausible but empirically underdeveloped cross-species comparisons. For example, it is claimed that a human and an octopus can both feel pain but their pains are realized by different neural mechanisms. Another traditional case for multiple realizability comes from thought experiments such as an alien species that feels pain but has a mental architecture that is not realized by a brain or organic compounds at all. Finally, many more recent arguments for multiple realization engage carefully with empirical research in order to find confirmation that psychological kinds can be realized by different neural kinds (e.g. Aizawa 2007; Richardson 2009; Figdor 2010). All strategies suggest that there are no coextensive mental and neural kinds because psychological kinds are multiple realized or at least multiple realizable.[9]

The multiple realization thesis is more than just an objection against coextensive kinds and often presented as a general argument against reductionism in philosophy of mind. It is uncontroversial that multiple realization arguments had a profound impact by making "non-reductive physicalism" a mainstream position in philosophy of mind. For example, Ernie LePore and Barry Loewer claim that it "is practically received wisdom among philosophers of mind that psychological properties (including content properties) are not identical to neurophysiological or other physical properties" (1989, 179). For much of the second half of the twentieth century, it was also rarely questioned that this "practically received wisdom" can be turned into a general argument against reductionism. As discussed in the previous sections, reductionism seems to require biconditional bridge principles that connect kinds on a vertical scale. If psychological kinds are multiple realized, however, there will no biconditional principles as mental kinds correspond to very different neural and physical kinds. Therefore, multiple realization undermines reductionism.

There are obvious similarities between the argument from multiple realization and the argument from horizontal pluralism. Both arguments share the premise that we will not always find coextensive kinds on a vertical scale. Furthermore, both arguments claim that reductionism fails because coextensive kinds are not available.

---

[9] Bickle (2003) distinguishes between multiple realization and multiple realizability. Furthermore, he declares "This broader sense of multiple realiz*ability* and philosophers' 'possible world' fantasies do not concern me. I don't know whether identity holds across 'all possible worlds,' or even across 'all physically possible worlds.' I don't know the 'conceptual' or 'nomological limits' of our psychological concepts." While I'm sympathetic with Bickle's metascientific methodology (see Chap. 11, footnote 1), it still seems to me that talk about realiz*ability* instead of realization can be perfectly legitimate in some areas of scientific practice. For example, research in biotechnology, engineering, and computer science can raise questions about multiple realiz*ability* that do not reduce to multiple realization. Even if ignore possible worlds and other toys of analytic metaphysics, there arguably remains a difference between what is multiple realized and what we could plausibly engineer in a multiple realized fashion. However, I will still follow Bickle in talking about multiple realization instead of realizability whenever possible.

This claim has to be qualified in the light of the discussion of the last section. The unavailability of coextensive kinds on a vertical scale challenges reductionism only if "reduction" is understood in an ambitious sense that contradicts vertical pluralism by aiming at global ontological and epistemic unification.

Despite these similarities, both arguments present different justifications for the claim that coextensive kinds are not available. While arguments from multiple realization typically rely on specific examples of allegedly multiple realized psychological kinds, the argument from horizontal pluralism states that a pluralism on the horizontal scale undermines the general expectation that every legitimate non-physical kind will correspond with a coextensive physical kind.

The differences between both arguments become of crucial importance when we consider contemporary criticisms of multiple realization. Much of the current discontent (cf. Bickle 2013) with multiple realization is motivated by two relates lines of arguments. First, it is pointed out that claims of multiple realization of mental states are often based on oversimplified accounts of neuroscientific research and that a closer look at contemporary neuroscience actually casts doubt on multiple realization. Second, it is suggested that many remaining and allegedly obvious cases of multiple realization are misunderstandings that stem from the use of overly coarse grained concepts that do not actually refer to legitimate scientific kinds.

The first line of argument is usually based on the observation that traditional cases of multiple realization of mental states rely on vague and rather superficial claims about neuroanatomical or neurophysiological differences in organisms with the same mental state M. However, these claims are surely not sufficient to show that M is multiply realized. Even under the assumption of some neuroanatomical and neurophysiological differences, we may still be able to identify shared neural mechanisms that realize M in essentially the same way. For example, Bechtel and Mundale's landmark paper (1999) discusses similarities in neural processing both within and across species. Considering a variety of examples such as neural processing of visual information, Bechtel and Mundale point out that much of cross-species research in cognitive neuroscience actually supports the assumption that there are important shared neural mechanisms that can explain psychological similarities.

Even if empirical evidence suggests that some psychological processes in mammals are not multiple realized, one can object that there are still countless obvious examples of multiple realization such as hunger in elephants and mosquitos, pain in humans and octopuses, or even learning in robots and humans. In response, critics of multiple realization often adopt a different strategy and doubt the legitimacy of a vague and general kind such as "hunger", "pain", or "learning". Maybe we only need domain-specific reductions such as "pain in humans" or "pain in molluscs" instead of a general account of "pain".

Critics of multiple realization can combine both strategies to create a dilemma for alleged cases of multiple realization: On the one hand, there are many psychological similarities that are based on neural similarities. In these cases, claims of multiple realization fail because there are no multiple realizations. On the other hand, cases without plausible neural similarities also raise doubts regarding psychological similarities. In these cases, claims of multiple realization fail because there are no robust psychological kinds that could be multiple realized.

Even if we accept current criticism of multiple realization in philosophy of mind, the argument from horizontal pluralism remains unchallenged. Dupré, for example, acknowledges that coextensive kinds can be found occasionally but insists that the rejection of essentialism undermines the belief that coextensive kinds can be found *everywhere*. Even if current developments in cognitive neuroscience undermine traditional appeals to multiple realization in philosophy of mind, reductionism as a general theory will still fail as they are countless scientific ontologies that include kinds that do not correspond to coextensive kinds on a supposedly more fundamental level.

For example, even a general failure of multiple realization in cognitive science would not undermine the claim that species such as *Taraxacum officinale* are multiply realized. Recall that the point of the argument from horizontal pluralism is not that we will *never* find coextensive kinds but rather that there are good reasons to assume we will not *always* find coextensive kinds. Given the diversity of ontologies that scientists can choose to work with, it is far from surprising that some ontologies include coextensive kinds on a vertical scale and others don't. For example, even if assume for the sake of the argument that some biological kinds are not multiply realized, there will also be perfectly legitimate species concepts "which have nothing to do with physical structure" (Dupré 1993, 105) such as an ecological species concept that defines species membership partly in terms of ecological niches.[10]

While evidence from cognitive neuroscience is therefore compatible with the argument from horizontal pluralism, some philosophers have extended their criticism of multiple realization claims beyond psychological kinds. Most importantly, Shapiro (2000, cf. Shapiro and Polger 2012) has challenged the entire framework of multiple realization arguments and its applications to debates about the "special sciences". Shapiro illustrates his argument with an example of corkscrews as an allegedly obvious case of a multiply realized kind. According to Shaprio, there are two lessons to learn from a discussion of this example. First, not every difference between instantiations of a kind is sufficient for multiple realization. For example, two realizations of a corkscrew that only differ in their color are not multiple realizations of corkscrews because "they do not differ in causally relevant properties – in properties that make a difference to how they contribute to the capacity under investigation" (Shapiro 2000, 644). Second, there remain examples of corkscrews with obvious differences in causally relevant properties such as instantiations of a waiter's corkscrew compared to instantiations of a winged corkscrew. However, Shapiro challenges the claim that *corkscrew* (or *mousetrap* or *eye* when applied to biological organisms and a camera) are legitimate kinds in a sense that requires meaningful generalizations or laws. Faced with the question what all instantiations of these kinds have in common, we can only answer with possibility

---

[10] Dupré's formulation "nothing to do with physical structure" is too strong, see footnote 7 and my discussion of "unique and interesting" similarities.

analytic and certainly "numbingly dull" laws such as "all mouse traps are used to catch mice" (Shapiro 2000, 649).

While Shapiro's dilemma – either a kind is not *multiply* realized or it is not a legitimate kind – may have intuitive force in cases such as corkscrews or mousetraps, it is highly implausible in the case of many examples from the empirical sciences including species. Take the ecological species concept and its application to common dandelions (*Taraxacum officinale*). If species membership is determined on the basis of shared ecological niches, there is little hope in finding physical kinds that distinguish members of a species from members of other species that inhabit slightly different ecological niches. However, this does not mean that we have only "numbingly dull" things to say about what members of an ecological species have in common. On the contrary, if we identify an organism as a member of a specific species, we gain a lot of highly relevant information about that organism. As mentioned in the last section, members of the species *Taraxacum officinale* share countless interesting properties such as being asexual, being genetically almost identical to their parent plants, being ruderal species, and so on.

Even if we accept current criticisms of multiple realization of psychological kinds, the general argument from horizontal pluralism remains unaffected as the example of species shows. There are countless biological kinds that match the explanatory interests of biologists and are certainly not "numbingly dull" but still have little to do with similarity on a physical level. However, one can also go a step further and argue that horizontal pluralism casts doubts on the claim that there are no multiple realizations in psychology. Instead, a horizontal pluralism regarding psychological ontologies seems to support the idea that multiply realization is plausible with regard to some but certainly not all psychological kinds.

To illustrate this claim, consider the vast diversity of psychiatric kinds that can be found in the latest editions of the *International Classification of Diseases* (ICD) or *Diagnostic and Statistical Manual of Mental Disorders* (DSM). Some psychiatric kinds are known to correspond with very specific neural phenomena that make the assumption of coextensive psychological and neural kinds attractive. Furthermore, neuropsychological discoveries about differences in neural processing are often accompanied by discoveries about corresponding psychological differences (cf. Soom et al. 2010). While all of this should make us suspicious about overly ambitious claims of multiple realization, it would also be very implausible to claim that every kind of the ICD or DSM comes with a coextensive neural kind (e.g. Schramme 2013). Clearly, there are also countless helpful psychiatric kinds that do not appear to correspond to interesting neural kinds. For example, Samuels (2009) argues that delusions do not correspond with coextensive neural kinds and therefore constitute irreducible cognitive kinds.

Of course, critics of multiple realization do not have to give up at this point but can employ Shapiro's dilemma to argue that psychiatric kinds without coextensive neural kinds should not be considered legitimate scientific kinds. But why not? Obviously, the argument needs to avoid circularity: we cannot argue that multiple realized psychiatric kinds are not legitimate kinds *because* they do not correspond to coextensive neural kinds. Still, one may suggest that psychiatric kinds such as delusions are merely folk-psychological kinds that will (or at least should) be eliminated from mature scientific psychiatry. Unfortunately, this is quite implausible as psychiatric kinds often fit nicely moderate accounts of natural kinds in terms of property clustering and projectibility (cf. Beebee and Sabbarton-Leary 2010; Ludwig 2015) and Samuels (2009) also develops the idea of delusions as homeostatic property clusters in detail.

As far as I can see, the only viable strategy to exclude all multiple realized psychiatric kinds would be based on a highly restrictive (e.g. essentialist) notion of natural kinds (cf. Haslam 2014). Psychiatric kinds without coextensive neural kinds may be useful in scientific practice, but they are not *genuine* natural kinds that "carve nature at its joints". Obviously, this strategy would be incompatible with my presentation of horizontal pluralism from the last chapter as I have argued that such a strong notion of natural kinds is not available in the life sciences. Even if successful, this strategy would therefore not undermine my claim that horizontal pluralism supports the assumption that we will not always find coextensive kinds on a vertical scale. Instead, this strategy would have to reject horizontal pluralism directly.

Let us consider another example of the link between horizontal and vertical pluralism in psychology by going back to debates about memory and extended cognition. Memory consolidation has become one of the most prominent examples in current debates about multiple realization and Bickle (2003, 148) has postulated a "cAMP–PKA–CREB molecular pathway, that uniquely realizes memory consolidation across biological classes, from insects to gastropods to mammals". Even if we accept his controversial (cf. Aizawa 2007; Sullivan 2008) claim for the sake of the argument, a horizontal pluralist will argue that we have little reason to assume that there are no multiple realized entities in memory research.

In Sect. 4.2 I argued for a plurality of legitimate accounts of memory that correspond with different explanatory interests in cognitive science. Cognitive scientists who are interested in neural mechanisms of memory will most likely opt for an internalist account of memory that makes claims of multiple realization less plausible and also lead to the empirical confirmation of interesting neural mechanisms that are shared by different species. While this restriction to biologically realized memory is perfectly fine in the context of certain research projects, cognitive scientists with different research interests (say computational modeling of problem solving behavior) will most likely opt for a different strategy. As discussed in the last chapter, a focus on problem solving behavior makes it attractive to use an externalist account of memory that accept extended realizations of memory (e.g. on screens and in notebooks) and therefore makes rejections of multiple realization highly implausible. A general rejection of multiple realization therefore seems to involve a biased sample and problematic circularity: while the examples are supposed to show that multiple realization is questionable, they are chosen *because* they are promising cases of coextensive psychological and neural kinds.[11]

---

[11] Of course, this problem of circularity applies in both directions. Philosophers, who reject coextensive psychological and neural kinds *tout court*, cannot appeal to their favorite examples of multiple realization, either. However, the point of the argument from horizontal pluralism is that

6.5   Reductive Explanation without Reduction

The argument from horizontal pluralism makes the case for irreducibility by assuming that reductions require coextensive scientific kinds on a vertical scale. However, horizontal pluralism gives us good reasons to believe that we will not always find coextensive kinds. Sometimes we will be able to identify kinds on a vertical scale, but there will also be perfectly legitimate scientific kinds that are not coextensive with kinds of the allegedly more fundamental level.

A vertical pluralism that rejects biconditional bridge principles and traditional theory reductions is hardly a novel position but a crucial aspect of many pluralist proposals in philosophy of science. Somewhat surprisingly, this pluralist mainstream in philosophy of science has largely failed to make an impact in philosophy of mind.[12] The limited interest of philosophers of mind in scientific pluralism is surprising because it seems to offer a clear challenge to traditional accounts of the mind-body problem. The idea that phenomenal consciousness constitutes a deep philosophical problem is based on the assumption that the explanatory gap between our phenomenal and physical accounts is unique and puzzling. However, vertical pluralism seems to challenge this idea by pointing out that explanatory gaps are ubiquitous in science. While it may be impossible to reduce phenomenal consciousness or intentionality, the situation is similar in the case of innocent biological entities such as common dandelions and their properties such as being an asexual or being a ruderal species. And if explanatory gaps are literally everywhere, there is no reason to be especially troubled by explanatory gaps in philosophy of mind.

Steven Horst (2007) is one of the few contemporary philosophers of mind who applies the lessons from pluralism in philosophy of science to the mind-body problem by arguing that "philosophy of mind [is] one of the last bastions 1950s philosophy of science" and that at the "entire problematic [of the 'hard problem of consciousness'] is an artifact of an erroneous view in the philosophy of science" (2007, 4). According to Horst, philosophy of mind is led astray by the outdated assumption that intertheoretic reductions are the norm in science and that failed reductions in philosophy of mind are mysterious exceptions. If the failure of intertheoretic reductions is ubiquitous in science, then irreducibility in philosophy of mind should not be considered a deep or unique problem.

Although the argument from horizontal pluralism seems to support Horst's claims, many philosophers will object that the problem of an explanatory gap in

---

we should expect the diversity of acceptable psychological ontologies to include psychological kinds that multiple realized as well as kinds that correspond to coextensive neural kinds.

[12] Still, there are numerous interesting pluralist proposals in philosophy of mind and cognitive science including van Bouwel (2014), Dale (2008), El-Hani and Pihlström (2002), Eronen (2011), Horst (2007), McCaunly and Bechtel (2001), Putnam (1999), Polger (2007), Schouten and Looren de Jong (2001). However, not all pluralist proposals will endorse a conceptual pluralism with both epistemic and ontological consequences. More specifically, many pluralists in contemporary philosophy of philosophy of neuroscience and cognitive science aim at an explanatory pluralism that avoids ontological issues or even insist on ontological unification.

philosophy of mind cannot be brushed aside that easily. Horst addresses this objection by discussing the intuition that "the psychological gaps are somehow different from and deeper than the others" (2007, 86). One way of justifying this intuition is to argue that we do not even have "candidate explainers" for consciousness while we have at least a broad idea of how a reductive story could look like in the case of biology and other special sciences. While Horst acknowledges the force of the intuition that there is something special about psychological gaps, he insists that the idea of candidate explainers fails in the light of pluralist philosophy of science. Again, the argument from horizontal pluralism can be used to back up this claim by suggesting that have do not even have "candidate explainers" for innocent entities such as species. Horst therefore argues that our intuitions "suffer from something like a Kantian dialectical illusion that leads us to the assumption that [all theories] can be reductively unified" (2007, 88).

Although I largely agree with Horst's diagnosis, most philosophers of mind will insist that it misses a crucial aspect of contemporary debates about reduction and reductive explanation. Indeed, the diagnosis that traditional accounts of reduction fail has become almost a truism but is often combined with the suggestion that a more moderate account of reductive explanation will still reveal an important difference between puzzling entities in philosophy of mind and innocent entities of the special sciences.[13] As Kim (2008, 94) puts it: "A certain picture seems widespread and influential in recent discussions of issues that involve reduction and reductive explanation—especially, in connection with the mind-body problem. The same picture is also influential in the way many think about the relationship between the 'higher-level' special sciences and 'basic' sciences. What I have in mind is the idea that *reducing* something is one thing and *reductively explaining* it is quite another."

The distinction between reduction and reductive explanation can be motivated through simple examples of ordinary entities such as "table." There can be no doubt that the ordinary kind "table" does not correspond to a coextensive and non-disjunctive physical kind as different tables have vastly different physical structures. While this observation suggests that we cannot reduce the ordinary kind table to a non-disjunctive physical kind, it still seems plausible that we can *explain* the macroproperties of a table in terms of the microproperties of the physical entities that compose the table. Consider a table and its properties such as a flat surface, a weight of 10 kg, four legs, and so on. Even if "table" is a multiply realized ordinary kind, it still seems obvious that the properties of every individual table can be physically explained. For example, the properties of having a flat surface and four legs are explained by the spatial arrangement of the physical entities that compose the table. The weight of the table is explained by the weight of the physical entities that

---

[13] Unsurprisingly, many reviews of Horst's *Beyond Reduction* stress this point. For example, Witmer (2010) warns us "The term 'reductionism' is nearly as bad as 'naturalism' in terms of being unclear and contested in its meaning. It may well be that the antireductionist consensus depends on a more restrictive notion of reduction than that had in mind by would-be reductionists in philosophy of mind. So just what is the sense of 'reduction' at issue here?" Brandon (2008) adds "that most scientists would think that the errors of philosophers show they have a bad model of reduction rather than that there are no reductions". See also my discussion in Sects. 6.5, 6.6, and 6.7.

compose the table. And so on. In the end, we have a physical explanation of every macro-property of the table and in this sense a reductive explanation of the table in terms ontologically prior microphysical entities. Therefore, there is no explanatory gap between our ordinary accounts of tables and physical accounts of the objects that constitute the table.

The example suggests a distinction between "reduction" and "reductive explanation." Reduction (in a philosophically ambitious sense that is incompatible with pluralism, cf. Sect. 6.3) requires coextensive kinds on a vertical scale, while reductive explanation can be successful without coextensive kinds and bridge principles. Furthermore, we can distinguish between "reductionism" as the claim that every true theory can be derived from a fundamental physical theory, and "reductivism" as the claim that every existing entity can be explained in terms of fundamental physical entities. The distinction between reductionism and reductivism casts doubt on pluralist positions that try to dissolve worries about explanatory gaps in philosophy of mind by pointing out the ubiquity of explanatory gaps in science. Contrary to Horst's (2007) claim that there is nothing special about explanatory gaps in philosophy of mind, reductivists can hold that phenomenal consciousness is indeed unique because of the failure of not only reductions but also reductive explanations.

It is not difficult to find prominent reductivist positions in contemporary philosophy of mind that combine the commitment to reductive explanation with a rejection of traditional accounts of theory reduction. Kim's framework of "functional reduction" (Kim 2005) offers a highly influential example that is often presented in direct opposition to the Nagelian model of theory reduction.[14] Functional reduction in the sense of Kim is a three-step process (e.g. Kim 2005, 101). The first step consists in a "functionalization" of a property in terms of a causal task C. The second step consists in finding properties or mechanisms in the reduction base that perform C. The last step requires an empirical theory that explains how the realizers of M perform C.

Kim illustrates his account of functional reduction with the example of genes. According to Kim, we can functionalize "gene" as the "mechanism that encodes and transmits genetic information" (2005, 101). Whatever it is that encodes and transmits genetic information will be a realizer of "gene". The second step consists in finding actual realizers in the reduction base and Kim points out that DNA molecules perform the task of encoding and transmitting genetic information. Finally, he suggests that molecular biology qualifies as the adequate empirical theory that explains how DNA molecules encode and transmit genetic information.

Kim's account of functional reduction comes with a number of noteworthy features. *First*, functional reduction is supposed to be compatible with multiple realization and with the argument from horizontal pluralism. Although the genes of all known life forms are realized by segments of DNA, there could also be a life form

---

[14] Marras (2002), however, raises important worries and argues that Kim's model of functional reduction essentially leads back to traditional accounts of theory reduction.

with a "mechanism that encodes and transmits genetic information" that is realized by non-organic material. Functional reduction therefore does not require coextensive kinds on a vertical scale and seems to escape the objections from the last section. *Second*, functional reductions are assumed to be applicable to entities in the so-called special sciences such as biology or psychology. My examples of the last section such as "being an asexual species", "having genetically identical parent plants", or "being ruderal species" can all be functionalized and therefore – at least in principle – functionally reduced. *Third*, phenomenal consciousness is an odd exception to this rule because we cannot functionalize phenomenal properties. While phenomenal properties such as pain come with typical causal roles, the phenomenal character is not captured by any functional analysis. *Fourth*, the general availability of reductive explanations and the odd exception of phenomenal consciousness reinforce the idea that the explanatory gap in philosophy of mind is somehow unique and mysterious.

Kim is by no means the only philosopher who subscribes to reductivism while rejecting reductionism. Another influential example is Chalmers who has developed a similar account under the changing labels of "logical supervenience" (1996), "reductive explanation" (Chalmers and Jackson 2001), and "scrutability" (2012). All three labels are used to develop the idea that we can explain almost all non-physical truths in terms of fundamental physical truths while phenomenal truths are odd and philosophically troubling exceptions. Chalmers' most recent (2012) scrutability formulation does not only resemble Kim's functional reduction in allowing multiple realization but even explicitly acknowledges that a "trend among philosophers of science has been to reject the strong unity of science thesis in favor of weaker theses such as connective theses or to argue that science is not unified at all" (2012, 301). Furthermore, Chalmers suggests that "scrutability is quite consistent with autonomy. If an economic truth (say about the financial crisis in 2008) is scrutable from physical truth, then a weak sort of explanation of the economic truths in terms of physical truths will be possible. [...] This 'explanation' may have little predictive power and little practical use. By contrast, an economic explanation may be far simpler, more systematic, more predictive, and more useful" (Chalmers 2012, 309). In other words: A weak reductivism in terms of scrutability will respect the crucial insights of contemporary philosophy of science while insisting on the unique status of inscrutable phenomenal truths.[15]

To sum up, the move from reduction to reductive explanation creates an important challenge for a pluralism that does not only make a claim about the

---

[15] This strategy is not only endorsed by philosophers of mind like Kim or Chalmers but similar issues occur in debates about "explanatory reduction" in biology (cf. Brigandt 2013; Kaiser 2011). For example, Weber (2005) argues for the compatibility of multiple realization and reductionism in biology by insisting that "the fact that chemotactic behavior is multiply realizable does not affect the reductionistic explanation of *this* organism's behavioral biology" (2005, 48). It will become clear in the next section, the same response is suggested by philosophers of mind and cognition who insist of reductive explanations but not on reductions of psychological states such as elementary learning. No matter what we think of reduction, we can explain elementary learning of *specific* organisms on a purely neural basis.

disunity of scientific practice but rejects the explanatory gap problem and the idea that we should be able to explain non-physical truths in terms of fundamental physical truths.

6.6 Reductive Explanations of Elementary Learning in *Aplysia*

Even if we accept the distinction between "reduction" and "reductive explanation", it is far from clear that we should accept a global reductivism. Furthermore, many pluralists will suspect that reductionism and reductivism are based on the same misunderstanding of scientific practice and misguided by reliance on highly questionable toy examples and thought experiments. A sober look at scientific practice does not give us any reason to believe that reductions or reductive explanations will be successful everywhere (e.g. Eronen 2011, 139).

Reductivists can reply that their position is actually motivated by case studies of reductive explanation in the empirical sciences. One intriguing example are reductive explanations in the psychology of learning. Consider examples of elementary learning such as habituation, sensitization, and classical conditioning. Habituation is a process in which a system learns that a stimulus is unimportant and ignores it. Sensitization is the converse of habituation, in which a system learns to recognize a stimulus as important and reacts to it an increasingly strong way. Classical conditioning describes a process in which a system learns to take a stimulus as an indicator of another upcoming stimulus and reacts accordingly.

There are many reasons to reject traditional theory reductions of learning psychology. Most obviously, learning psychology does not come with laws akin to those in classical cases of theory reduction. Furthermore, psychological kinds such as habituation, sensitization, or classical conditioning can be realized in a large variety of vastly different organisms such as mammals and molluscs. This observations makes it *prima facie* plausible to assume that kinds in learning psychology are multiple realized and therefore not reducible.

While the multiple realization of elementary learning processes such as habituation may be *prima facie* plausible in the light of neuroanatomical differences between mammals and molluscs, it has become challenged on the basis of a more careful engagement with molecular neuroscience. Bickle (2003) presents memory consolidation as one of his major examples for reductionistic neuroscience and explicitly considers the case of *Aplysia*. According to Bickle, molecular neuroscience does not only reveal shared mechanisms in memory consolidation of invertebrates such as *Aplysia* and *Drosophila* but he argues for a "cAMP–PKA–CREB molecular pathway, that uniquely realizes memory consolidation across biological classes, from insects to gastropods to mammals" (2003, 132).

Aizawa challenges Bickle's claim on the basis of a discussion of PKA (protein kinase A) and CREB (cAMP response element binding proteins) that constitute crucial elements of Bickle's postulated cross-species molecular pathway of memory consolidation. Aizawa's main point is that neither PKA nor CREB are the same in invertebrates such as *Aplysia* and mammals such as humans. As Aizawa puts it: "if there are biochemical pathways $cAMP–PKA_1–CREB$, $cAMP–PKA_2–CREB$, and $cAMP–PKA_3–CREB$, where $PKA_1 = PKA_2 = PKA_3$, then no matter what other elements might be added to the sequence – elements such as the ubiquitin hydrolase mentioned above – the pathway is still multiply realized" (Aizawa 2007, 68).

We do not need a final decision of this debate about multiple realization of memory consolidation to see why reductive explanations in learning psychology seem attractive independently of the fate of reductions. Even if reductions of learning psychology fail, it still seems plausible that one can reductively explain learning in simple organisms such as *Aplysia*. The case for reductive explanations without reductions in learning psychology is pretty straightforward and does not even require detailed engagement with the empirical state of play but can be illustrated on the basis of the textbook case of Eric Kandel et al.'s groundbreaking research on learning in *Aplysia* (e.g. Carew et al. 1981; cf. Kandel 2006).[16] The structure of its nervous system makes *Aplysia* an extraordinarily interesting model organism. The brain of *Aplysia* is composed of 20,000 cells and nine separate clusters, the so called ganglia. Therefore, a single ganglion is composed of roughly 2000 cells, which is an astonishingly small number when compared to the roughly 100 billion cells that comprise the human brain. The small number of neurons combined with their unusually large size makes *Aplysia* a promising model organism for neuroscientists.

At the same time, *Aplysia* is capable of learning and all three of the elementary learning mechanisms mentioned above can be found in *Aplysia* (Carew et al. 1981). Kandel et al. tested *Aplysia*'s learning mechanisms with experiments involving the siphon, a tube-like organ which is common among sea snails. Through habituation *Aplysia* learns to ignore an irrelevant stimulus. When the experimenter touches the siphon, the gills withdraw as a protective reflex. However, this reflex is not invariant. If touching the siphon remains without consequences, the responses of *Aplysia* become increasingly weaker. Sensitization involves the opposite learning process; if touching the siphon is primed by an electric shock, the reaction becomes increasingly stronger. In the classical conditioning task, touching the siphon is used as an indicator of an upcoming electric shock and the protective gill-withdrawal reflex is correspondingly strong.

Kandel's research on *Aplysia* is not limited to learning psychology, but also focuses on the neuroscience of learning in *Aplysia*. Already in 1968 his group found six motor neurons that cause the gill-withdrawal reflex. Six sensory neurons are connected to these motor neurons, allowing Kandel to describe a neuronal correlate of the gill withdrawal reflex: "[T]he neural architecture of at least one behavior of Aplysia was amazingly precise. In time, we found the same specificity and invariance in the neural circuitry of other behaviors" Kandel (2006, 196).

---

[16] For more recent and careful discussions, see review articles by Rankin et al. (2009) and Glanzman (2006).

The possibility of learning despite invariant neural connections is explained by an idea that was earlier formulated by Jerzy Konorski (1950) and Donald Hebb (1949). This is that learning need not be realized by new connections; rather it is sufficient if the strength of connections is variable. In the case of *Aplysia*, Kandel was able to specify the idea of the "strength of connections" with a precise molecular phenomenon: the amount of the neurotransmitter glutamate. "During short-term habituation lasting minutes, the sensory neuron releases less neurotransmitter and during short-term sensitization it releases more neurotransmitter" (Kandel 2006, 222).

The molecular description of the neurophysiological processes allows us not only to derive *Aplysia's* reaction to the stimuli, but also to derive the behavioral change of *Aplysia* given a series of stimuli. In the case of habituation, we understand on a molecular level why the gill-withdrawal reflex becomes increasingly weaker. Touching the siphon stimulates the sensory neurons, which are connected to six motor neurons, which cause the gill-withdrawal reflex. The synaptic transmission from sensory neurons to motor neurons is based on the release of the neurotransmitter glutamate, and habituation in *Aplysia* works through the decrease of the amount of glutamate released by the sensory neuron.

Given Kandel's description of the molecular processes, it seems highly plausible that we have a reductive explanation of elementary learning processes such as habituation in *Aplysia*. Habituation in Aplysia *is nothing but* the molecular process described by Kandel. If someone would insist that Kandel's theory explains molecular processes but not habituation, we could rightly object that this person is confused about the meaning of "habituation." "Habituation" is nothing but a behavioral pattern and disposition in which a system increasingly ignores stimuli. Therefore, Kandel's molecular theory explains habituation in *Aplysia*. Or, to put it in terms of Kim's model of functional reduction: First, we can functionalize habituation as a causal task of a decreased response to a stimulus under certain boundary conditions.[17] Second, we can identify the neurophysiological processes that realizes the decreased response to a stimulus in *Aplysia*. Third, we can argue that Kandel's research provides an adequate empirical theory that explains how the stimulus-response is decreased in *Aplysia*.

It seems that the case of *Aplysia* provides an attractive example of the availability of reductive explanations independently of the messy issues of theory reduction. No matter whether theories of elementary learning can be reduced to neuroscientific theories, it seems plausible that we can explain instantiations of elementary learning such as habituation, sensitization, and classical conditioning in terms of neural processes. We do not even have to assume the availability of "local reductions" in the sense of Kim (1993). Imagine that the described neural processes can only be

---

[17] Gold and Stoljar (1999) have argued that reductions of elementary learning to biological neuroscience fail because Kandel's account of elementary learning in Aplysia is based on psychological theory and therefore does not count as "purely biological". Their discussion is helpful in illustrating that the functionalization of elementary learning will depend on psychological theory. However, a reductivist will point out that this dependency is compatible with a more moderate account of reductive explanation that does not aim at traditional theory reduction.

found in one individual member of *Aplysia* while all other members of *Aplysia* realize elementary learning in a different way. It would still be plausible to say that we have a reductive explanation of elementary learning in that individual member of *Aplysia*.

The case of *Aplysia* illustrates why the distinction between reduction and reductive explanation challenges the project of a pluralist theory of the mind. Recall my claim that we do not have to worry about the irreducibility of the mind. It now seems that this is true with respect to theory reduction but not with respect to reductive explanation. The argument from horizontal pluralism justifies the assumption that mental kinds cannot be reduced to coextensive neural kinds but it still seems reasonable to assume that there must be a reductive explanation for every individual mental state.

Given this differentiation between reduction and reductive explanation, the argument from horizontal pluralism seems unsuccessful in deflating common formulations of the mind-body problem. Consider phenomenal consciousness and the question of whether there is a physical explanation of phenomenal consciousness. Here, the problem is not the unavailability of coextensive kinds but the apparent unavailability of reductive explanations. Reductivists will insist that it is this unexplained phenomenal side of the mental and not multiple realization that constitutes the core of the mind-body problem.

In the following sections, I will challenge reductivism and argue that a thorough pluralism should reject the universal availability of both theory reductions and reductive explanations. In next Sect. 6.7, I will argue that the separation of reduction and reductive explanation comes with the unacceptable price of eliminativism towards non-physical properties. In the following Chap. 7 I will argue that even if we accept that reductivism is clearly separated from reductionism, there is still no reason to believe that reductivism is true.

6.7   Limits of Reductivism

A crucial advantage of reductivism compared to reductionism is its compatibility with multiple realization and with the unavailability of coextensive kinds on a vertical scale. While this compatibility of reductivism and multiple realization challenges pluralist frameworks, a closer look at different ontologies on a vertical scale raises doubts that this challenge will be successful.

One important problem of reductivism is based on the question what instantiations of a reductively explained kind have in common. A popular answer to this question is based on the idea of local reduction. In the case of elementary learning, there are only species-specific kinds such as classical-conditioning-in-*Aplysia*, classical-conditioning-in-humans, and so on. We do not need a reductive account of classical conditioning across different species because "classical conditioning" is far too coarse-grained to constitute an interesting scientific kind.

Unfortunately, this strategy will not be successful with regard to all scientific kinds. For example, recall my example of species such as the common dandelion *Taraxacum officinale* and the claim that species do not have coextensive physical kinds. While it seems plausible to replace classical conditioning with more specific kinds such as classical-conditioning-in-*Aplysia*, it is far less attractive to replace species with more specific kinds such as subspecies. Furthermore, even if we would replace *Taraxacum officinale* with more specific subspecies, there would be still be no reason to believe that these more specific kinds would correspond with coextensive physical kinds.

A reductivist may respond by arguing that there is actually a simple answer to the question what reductively explained instantiations of a kind have in common: they share a functional role (cf. Kim 2005; Marras 2002). This answer immediately raises the question what it means to share a functional role. If reductivism is supposed to be compatible with the multiple realization of scientific kinds such as *Taraxacum officinale*, a shared functional role cannot be explained in terms of a shared non-disjunctive physical property. Kim (2008, 109) freely admits that this situation puts reductivists in the uncomfortable position of having to choose between what he calls "functional property realism" and "functional property conceptualism".

The realist option states that a functional role refers to a functional property that cannot be reduced to lower-level physical properties. For example, asexual organisms share the functional property of being asexual which is distinct from any physical property. Kim rejects functional property realism by arguing that it is an "essentially antireductionist and antiphysicalist view" (2008, 110). Indeed, the postulation of ontologically autonomous higher-level (e.g. biological, psychological, and social) properties would imply vertical pluralism and undermine the goal of global ontological and epistemic unification. If reductivism entails functional property realism, it is no treat to vertical pluralism.

Kim therefore suggests that we have to endorse functional property conceptualism. According to this view, instantiations of a functional property only share that they fall under the same concept. For example, there isn't any property that is shared by all (and only by) asexual organisms and they only share that they fall under the concept "asexual organism". The obvious difficulty with this conceptualist approach is that it implies an eliminativist stance towards functional properties. Functional properties cannot be reduced to physical properties and they are not ontologically autonomous, either. We are therefore left without functional properties and have to content ourselves with functional predicates. While Kim is clearly not happy with this implication, he considers all other options even less attractive and concludes that "we may well have to live with mental eliminativism and irrealism" (Kim 2008, 112).

I think that Kim is far too quick in accepting this kind of eliminativism as a reasonable option. First, the problem is not unique to psychological properties and therefore does not only imply a *mental* eliminativism. Instead, it occurs with virtually all properties of the so-called special sciences and would also imply a highly implausible *biological* eliminativism and irrealism. Consider the case of species as discussed in the last sections. What have different instantiations of one species – e.g. common dandelions or jaguars – in common? As discussed in Sect. 6.2, there is no reason to believe that they have unique non-disjunctive physical properties in common. In fact, it seems largely uncontroversial that there is no unique and interesting physical property that corresponds with species membership. Given the unavailability of corresponding physical properties and Kim's rejection of functional property realism, he has to endorse an eliminativist position according to which the only thing that common dandelions or jaguars have in common is that they fall under the concept "common dandelion" or "jaguar".

However, this is staggeringly implausible as members of the same species share *a lot* of interesting anatomical, behavioral, ecological, genetic, morphological, and phylogenetic features. If we identify an individual as a member of a certain species, we gain a lot information that challenges the conceptualist's claim that species membership only amounts to falling under a shared concept. Instead, it seems almost trivial to claim that species concepts are important in biology because they capture interesting patterns in nature. Functional property conceptualism would imply an extreme form of conventionalism as we would not be able to say anything substantive about the shared nature of members of the same species.

The implausibility of such a position becomes even clearer in the light of my discussions of natural kinds and realism in Sects. 4.4 and 4.5. At first, one may suspect that species pluralism actually supports a functional property conceptualism: if we can choose between different species concepts, doesn't species membership reduce to falling under the same conventionally constructed concept? However, I have argued that even a staunch pluralism requires more than conventions as the choice of a species concept is not random but motivated by biological (e.g. anatomical, ecological, genetic, phylogenetic) similarities that are interesting for biological research. As I discussed in the section on natural kinds, these empirically discovered patterns usually build the basis for more moderate account of natural kinds as they reflect the contingent clustering properties and their value for inductive inference. Far from collapsing scientific kinds into "anything goes"-conventions, the pluralism of the last chapter does not support a conceptualism that states that members of a biological kind only share falling under the same concept.

One way of further motivating such a moderate realism comes from ethnobiology and research on 1:1 correspondences between generic taxa in indigenous taxonomies and modern scientific taxa (cf. Ludwig 2015 for an extensive discussion). Hunn and Brown (2011), for example, report correspondences between 44 and 92 % between Tzeltal animal generics in southern Mexico and species taxa in contemporary biology. According to Berlin et al. (1974, 102), 61 % of Tzeltal plant generics correspond perfectly to scientific species. A pluralist approach can easily explain the differences between indigenous and modern western taxa by pointing out that taxonomies are shaped by different interests and criteria. For example, most indigenous cultures distinguish between organisms on the basis ecological criteria that play little or no role in modern biological taxonomies that largely rely on phylogenetic considerations.

Still, there is considerable cross-cultural agreement that needs to be explained and suggests a shared biological reality that contains what Atran and Medin call a "world of fairly stable clusters of complex features, whose remarkable endurance in the face of constant change can presumably be explained in terms of naturally occurring causal patterns." (Atran and Medin 2008, 11). Consider jaguars as one of the countless examples of 1:1 taxonomic correspondence between modern biological species (*Panthera onca*) and generic taxa in indigenous taxonomies of Central America (e.g. *Chak Mo'ol, Báalam, Chak bolay*, see Schlesinger 2001). This observation of widespread cross-cultural agreement seems everything but surprising simply because jaguars clearly differ from any other species in the Americas despite some similarities to pumas, ocelots, and margays. Different taxonomies may weight anatomical, behavioral, ecological, genetic, morphological, phylogenetic properties differently but these weightings will not lead to different taxa in this case because jaguars constitute a clearly identifiable cluster on each of these levels.

It is not difficult to see how the assumption of biological patterns challenges a reductivism in the spirit of Kim. As argued in the last sections, biological kinds such as species do not correspond to non-disjunctive physical kinds. The question what members of a taxon have in common therefore cannot be answered on a purely physical level. Kim acknowledges that this situation leaves reductivists with two options: functional property realism or functional property conceptualism. As the realist option would imply a pluralism of countless ontologically autonomous biological, psychological, and social entities, he opts for the conceptualist option. This conceptualism implies an eliminativism that may have some *prima facie* plausibility in the case of some folk-psychological kinds, but fails to offer any account of many other scientific kinds that clearly refer to biological patterns.

It is important to note that this problem is by no means unique to Kim's account but rather a challenge for reductivism in general. Consider, for example, Chalmers' most recent account of "a priori scrutability" (2012) that fits my characterization of reductivism as it is presented as "a sort of reductive explanation without reduction" (2012, 309) that avoids classical problems of theory reduction by claiming that non-fundamental (e.g. biological) truths are scrutable from fundamental (e.g. microphysical) truths without requiring any form of theory reduction. Chalmers suggests a scrutability base PQTI that contains four elements: physical truths (P), phenomenal truths (Q), indexical truths (I), and a negative "that's all" clause (T). Given a sufficient understanding of non-fundamental truths that is granted by conceptual analysis, PQTI would put us in position to know every other truth a priori.

It is not difficult to see why Chalmers' scrutibility framework runs into the same problems as Kim's account of functional reduction. Chalmers argues that his account is compatible with multiple realization. For example, truths about biological similarities do not require corresponding truths about physical similarities. Two individuals may be similar in being asexual organisms or even belonging to the same species without differing in any interesting way from sexual organisms or members of other species on a physical level. Instead, Chalmers assumes that a sufficient understanding of the concepts such as "asexual organism" or "common dandelion" will allow us to explain biological truths in terms of physical truths even if members of the same species do not share unique and interesting physical properties. In other words: conceptual analysis ensures that biological truths are a priori scrutable from PQTI. However, this suggestion leads straight back to the problem of eliminativism as it locates biological similarity solely in the conceptual realm. For example, we cannot distinguish biological kinds from gerrymandered kinds that arbitrarily group together objects with vastly different physical structures. In order to avoid this eliminativism, we would have to take biological similarities and patterns seriously and argue they do not reduce to falling under the same concept.

All of this clearly does not challenge the value of reductive explanations in scientific practice. Neither does it challenge a generalized account of reductive explanation that is compatible with vertical pluralism by endorsing what Kim calls "functional property realism". However, it undermines variants of reductivism that contradict vertical pluralism by aiming at global ontological and epistemic unification.

Let us take stock. The aim of this chapter was to evaluate whether horizontal pluralism gives us any reason to endorse vertical pluralism. I have presented an "argument from horizontal pluralism" according to which the availability of a plurality of equally legitimate ontologies on the horizontal scale undermines the availability of matching ontologies with coextensive kinds on a vertical scale. One strength of this argument is that it is not affected by criticism of multiple realization that challenges common examples of multiple realization in cognitive neuroscience. Even if we accept this criticism, the argument from horizontal pluralism still suggests that we will not always find coextensive kinds. However, even if this undermines traditional accounts of reduction, many philosophers of mind will insist that it does not provide any reason to doubt the widespread availability of reductive explanations. Philosophers like Chalmers (1996), Levine (2001), and Kim (2005) use this reductivist response to argue for the uniqueness of an explanatory gap regarding consciousness and therefore challenge pluralist accounts in philosophy of mind.

I have objected that this strategy of reductive explanations without reductions either leads to an ontological pluralism on the vertical scale or implies an unacceptable eliminativism regarding scientific entities. In the case of Kim, the model of "functional reduction" entails a functional property conceptualism that leaves no room for biological and other "higher level" properties. For example, Kim's account implies that the only thing that members of a species have in common is that they fall under a species concept and does not even allow a minimal realism about biological patterns that are constituted by anatomical, behavioral, ecological, genetic, morphological, phylogenetic similarities. Chalmers' scrutability framework ends up with the same problems as he insists that all truths are a priori scrutable from PQTI. Given that biological kinds do not correspond to coextensive physical kinds, Chalmers' account implies that the question what instantiations of a biological pattern have in common has no substantive answer that goes beyond them falling under the same concept.

Of course, reductivists could bite the bullet and accept eliminativism regarding biological entities such as species. However, it seems highly implausible that species membership reduces to falling under the same concept. If we identify an organism as a member of a certain species, we gain a lot of information about this organism that also comes with considerable predictive power. It would be very odd to claim that the only thing that members of the same species have in common is that they fall under the same concept. Furthermore, this eliminativism is also in direct conflict with a "naturalism of scientific practice" (see Chap. 1). According to this brand of naturalism, philosophers do not have a special grasp of the fundamental structure of reality that would allow them to take a general revisionist stance towards the ontologies of the empirical sciences. In other words: philosophers are not in an epistemic position to defend a "functional property conceptualism" that tells scientists that most of their entities do not really exist. Instead, philosophers should aim at metaphysical accounts that take the existence claims of the empirical sciences seriously.

References

Aizawa, Kenneth. 2007. The Biochemistry of Memory Consolidation: A Model System for the Philosophy of Mind. *Synthese* 155 (1): 65–98.

Atran, Scott, and Douglas L. Medin. 2008. *The Native Mind and the Cultural Construction of Nature*. Cambridge, Mass.: MIT Press.

Bapteste, Eric, and John Dupré. 2013. Towards a processual microbial ontology. *Biology & Philosophy* 28 (2): 379–404.

Bechtel, William, and Jennifer Mundale. 1999. Multiple Realizability Revisited: Linking Cognitive and Neural States. *Philosophy of Science* 66 (2): 175–207.

Beebee, Helen, and Nigel Sabbarton-Leary. 2010. Are Psychiatric Kinds Real?. *European Journal of Analytic Philosophy* 6 (1): 11–27.

Berlin, Brent, Dennis Eugene Breedlove, and Peter H. Raven. 1974. Principles of Tzeltal Plant Classification. *Human Ecology* 5 (2): 171–75.

Bickle, John. 2003. Philosophy and Neuroscience: *A Ruthlessly Reductive Account*. Boston: Kluwer.

Bickle, John. 2013. Multiple Realizability. In *The Stanford Encyclopedia of Philosophy*, edited by Edward N. Zalta, Spring 2013. http://plato.stanford.edu/entries/multiple-realizability/

Van Bouwel, Jeroen. 2014. Pluralists about pluralism? Different versions of explanatory pluralism in psychiatry. *New Directions in the Philosophy of Science,* eds. Galavotti et al., 105–119, Berlin: Springer.

Brandon, Ed. 2008. Review – Beyond Reduction. *Metapsychology Online Reviews*.

Brigandt, Ingo. 2013. Explanation in Biology: Reduction, Pluralism, and Explanatory Aims. *Science & Education* 22 (1): 69–91.

Brigandt, Ingo and Love, Alan. 2012 Reductionism in Biology, *The Stanford Encyclopedia of Philosophy*, http://plato.stanford.edu/archives/sum2012/entries/reduction-biology

Carew, Thomas, Edgar Walters, and Eric Kandel. 1981. Classical Conditioning in a Simple Withdrawal Reflex in Aplysia Californica. *The Journal of Neuroscience* 1 (12): 1426–37.

Chalmers, David. 1996. *The Conscious Mind: In Search of a Fundamental Theory*. Oxford: Oxford University Press.

Chalmers, David. 2012. *Constructing the World*. Oxford: Oxford University Press.

Chalmers, David J., and Frank Jackson. 2001. Conceptual Analysis and Reductive Explanation. *Philosophical Review* 110: 315–60.

Chang, Hasok. 2012. *Is Water H2O?: Evidence, Realism and Pluralism*. Berlin: Springer.

Clarke, Ellen. 2013. The multiple realizability of biological individuals. *The Journal of Philosophy* 110 (8): 413–435.

Craver, Carl F. 2007. *Explaining the Brain*. Oxford: Oxford University Press.

Dale, Rick. 2008. The Possibility of a Pluralist Cognitive Science. *Journal of Experimental and Theoretical Artificial Intelligence* 20 (3): 155–179.

Devitt, Michael. 2008. Resurrecting Biological Essentialism. *Philosophy of Science* 75, 344–82.

Dizadji-Bahmani, Foad, Roman Frigg, and Stephan Hartmann. 2010. Who's Afraid of Nagelian Reduction?. *Erkenntnis* 73 (3): 393–412.

Dupré, John. 1993. *The Disorder of Things: Metaphysical Foundations of the Disunity of Science*. Cambridge, Mass.: Harvard University Press.

El-Hani, Charbel Niño and Sami Pihlström. 2002. Emergence Theories and Pragmatic Realism. *Essays in Philosophy* 3 (2): 1–40.

Ereshefsky, Marc. 2010. What's wrong with the new biological essentialism. *Philosophy of Science* 77 (5): 674–685.

Eronen, Markus I. 2011. *Reduction in Philosophy of Mind: A Pluralistic Account*. Berlin: Walter de Gruyter.

Eronen, Markus I. 2013. No Levels, No Problems: Downward Causation in Neuroscience. *Philosophy of Science* 80(5): 1042–1052.

Figdor, Carrie. 2010. Neuroscience and the Multiple Realization of Cognitive Functions. *Philosophy of Science* 77 (3): 419–456.

Glanzman, David L. 2006. The Cellular Mechanisms of Learning in Aplysia: of Blind Men and Elephants. *The Biological Bulletin* 210 (3): 271–279.

Gold, Ian, and Daniel Stoljar. 1999. A neuron doctrine in the philosophy of neuroscience. *Behavioral and Brain Sciences* 22 (5): 809–830.

Haslam, Nick. 2014. Natural Kinds in Psychiatry: Conceptually Implausible, Empirically Questionable, and Stigmatizing. *Classifying Psychopathology: Mental Kinds and Natural Kinds,* eds., Harold Kincaid and Jacqueline A Sullivan, 11–28. Cambridge, Mass.: MIT Press.

Hebb, Donald Olding. 1949. *The Organisation of Behaviour*. New York: Wiley.

Horst, Steven W. 2007. *Beyond Reduction: Philosophy of Mind and Post-reductionist Philosophy of Science*. Oxford: Oxford University Press.

Hunn, Eugene S., and Cecil H. Brown. 2011. Linguistic Ethnobiology. In *Ethnobiology: What Can We Learn About the Mind as Well as Human Environmental Interaction?*, 319–33. New York: Wiley.

Kaiser, Marie I. 2011. The Limits of Reductionism in the Life Sciences. *History & Philosophy of the Life Sciences* 33 (4): 453–476.

Kaiser, Marie I. 2012. Why it is Time to Move Beyond Nagelian Reduction. *Probabilities, Laws, and Structures*, eds. Dennis Dieks, Wenceslao J. Gonzalez, Stephan Hartmann et al., 245–262. Berlin: Springer.

Kandel, Eric R. 2006. *In Search of Memory: The Emergence of a New Science of Mind*. New York: W. W. Norton & Company.

Kim, Jaegwon. 1993. *Supervenience and Mind: Selected Philosophical Essays*. Cambridge: Cambridge University Press.

Kim, Jaegwon. 2005. *Physicalism, or Something Near Enough*. Princeton: Princeton University Press.

Kim, Jaegwon. 2008. Reduction and Reductive Explanation: Is One Possible Without the Other. In *Being Reduced*, eds. Jakob Hohwy and Jesper Kallestrup, 93–114, Oxford: Oxford University Press.

Konorski, Jerzy. 1950. Mechanisms of Learning. In *Symposia of the Society for Experimental Biology IV* 4: 409–31.

Ladyman, James, and Don Ross,. 2007. *Every Thing Must Go: Metaphysics Naturalized*. Oxford: Oxford University Press.

LaPorte, Joseph. 2004. *Natural Kinds and Conceptual Change*. Cambridge: Cambridge University Press.

Levine, Joseph. 2001. *Purple Haze: The Puzzle of Consciousness*. Oxford: Oxford University Press.

LePore, Ernest, and Barry Loewer. 1989. More on Making Mind Matter in Philosophy of Mind. *Philosophical Topics* 17 (1): 175–91.

Ludwig, David. 2015. Indigenous and Scientific Kinds. *The British Journal of the Philosophy of Science, first published online*.

Lynch, Michael P. 2001. *Truth in Context: An Essay on Pluralism and Objectivity*. Cambridge, Mass.: MIT Press.

Marras, Ausonio. 2002. Kim on Reduction. *Erkenntnis* 57 (2): 231–57.

McCauley, Robert N., and William Bechtel. 2001. Explanatory Pluralism and Heuristic Identity Theory. *Theory and Psychology* 11 (6): 736–760.

Mitchell, Sandra D. 2003. *Biological Complexity and Integrative Pluralism*. Cambridge: Cambridge University Press.

Nagel, Ernest. 1979. *The Structure of Science: Problems in the Logic of Scientific Explanation*. Indianapolis: Hackett.

Okasha, Samir. 2002. Darwinian Metaphysics: Species and the Question of Essentialism. *Synthese* 131: 191–213.

Polger, Thomas W. 2007. Some Metaphysical Anxieties of Reductionism. *The Matter of the Mind: Philosophical Essays on Psychology, Neuroscience and Reduction*, eds. Maurice Kenneth Davy Schouten, and Huibert Looren de Jong, 51–75. Cambridge, Mass.: MIT Press.

Price, Huw. 1992. Metaphysical Pluralism. *The Journal of Philosophy* 89 (8): 387–409.

Putnam, Hilary. 1988. *Representation and Reality*. Cambridge: Cambridge University Press.

Putnam, Hilary. 1999. *The Threefold Cord: Mind, Body, and World*. New York: Columbia University Press.

Putnam, Hilary. 2004. *Ethics Without Ontology*. Harvard: Harvard University Press.

Rankin, Catharine, et al. 2009. Habituation Revisited: an Updated and Revised Description of the Behavioral Characteristics of Habituation. *Neurobiology of Learning and Memory* 92 (2): 135–138.

Richardson, Robert C. 1979. Functionalism and Reductionism. *Philosophy of Science* 46 (4): 533–58.

Richardson, Robert C. 2009. Multiple Realization and Methodological Pluralism. *Synthese* 167 (3): 473–92.

Samuels, Richard. 2009. Delusion as a natural kind. In *Psychiatry as cognitive neuroscience: Philosophical perspectives*, eds. Matthew R Broome and Lisa Bortolotti, 49–79. Oxford: Oxford University Press.

Schaffner, Kenneth F. 1967. Approaches to Reduction. *Philosophy of Science* 34: 137–147.

Schaffner, Kenneth F. 1976. Reductionism in Biology: Prospects and Problems. In *PSA 1974*: 613–632.

Schlesinger, Victoria. 2001. *Animals and Plants of the Ancient Maya: A Guide*. Austin: University of Texas Press.

Schramme, Thomas. 2013. On the autonomy of the concept of disease in psychiatry. *Frontiers in psychology* 4: 1–9.

Shapiro, Lawrence A. 2000. Multiple Realizations. *The Journal of Philosophy* 97: 635–54.

Shapiro, Lawrence A., and Thomas W. Polger. 2012. Identity, Variability, and Multiple Realization in the Special Sciences. In *New Perspectives on Type Identity: The Mental and the Physical*, eds. Simone Gozzano and Christopher S Hill, 264–88. Cambridge: Cambridge University Press.

Schouten, Maurice, and Looren de Jong, Huib. 2001. Pluralism and Heuristic Identification Some Explorations in Behavioral Genetics. *Theory and Psychology* 11 (6): 796–807.

Soom, Patrice, Christian Sachse, and Michael Esfeld. 2010. Psycho-neural Reduction Through Functional Sub-Types. *Journal of Consciousness Studies* 17 (1): 7–26.

Sullivan, Jacqueline Anne. 2008. Memory Consolidation, Multiple Realizations, and Modest Reductions. *Philosophy of Science* 75 (5): 501–513.

Walter, Sven. 2006. Multiple Realizability and Reduction: A Defense of the Disjunctive Move. *Metaphysica* 9: 43–65.

Weber, Marcel. 2005. *Philosophy of experimental biology.* Cambridge: Cambridge University Press

Wimsatt, William C. 1976. Reductionism, levels of organization, and the mind-body problem. *Consciousness and the brain,* eds. Gordon G Globus et al., 205–267. Berlin: Springer.

Witmer, Gene. 2010. Review Steven Horst Beyond Reduction: Philosophy of Mind and Post-Reductionist Philosophy of Science *Notre Dame Philosophical Reviews.*

Chapter 7
The Argument from Ontological Non-fundamentalism

I have sketched the strategy of a pluralist theory of the mind along the following lines: we do not need to worry about the irreducibility of the mind because there is nothing mysterious about irreducibility in general. While reductions often play an important role in scientific practice, there is no good reason to believe that *everything* must be reducible to a fundamental ontology. In fact, the argument from horizontal pluralism suggests that irreducibility extends to innocent entities such as species in our most trusted scientific disciplines such as biology. The most common way of introducing the mind-body problem is to point out the irreducibility of the mind. However, if irreducibility is common and unproblematic in scientific practice, there is nothing mysterious about the irreducibility of the mind and the mind-body problem vanishes in the light of a more realistic account of scientific practice.

Philosophers of mind like Chalmers, Jackson, Levine, and Kim will not be impressed by this argument because they employ a distinction between theory reduction and reductive explanation: even if there are limits to traditional theory reductions, we should still expect reductive explanations of mental states. The argument from horizontal pluralism might offer good reasons to reject reductionism in the sense of generalized theory reduction, but it does not offer any reason to reject reductivism in the sense of generalized reductive explanations. At this point, philosophers of mind can insist that the core of the mind-body problem is not irreducibility, but the lack of reductive explanations of mental states (or, more specifically, phenomenal states). Vertical pluralism as it is implied by the argument from horizontal pluralism does not justify the claim that we do not need a reductive explanation of mental states.

In the last section, I argued that this reductivist strategy comes at the unacceptably high price of an eliminativism regarding entities in disciplines such as biology or cognitive science. For example, we cannot reductively explain truths about biological kinds unless we claim that kind membership reduces to falling under the same concept. In this section, I want to present a second argument against reductivism that is based on the assumption that there is an even more direct connection between conceptual relativity and vertical pluralism. Conceptual relativity challenges the

© Springer International Publishing Switzerland 2015
D. Ludwig, *A Pluralist Theory of the Mind*, European Studies
in Philosophy of Science 2, DOI 10.1007/978-3-319-22738-2_7

ideal of exactly one fundamental scientific ontology. In the following, I will present an "argument from ontological non-fundamentalism" according to which this rejection of exactly one fundamental ontology undermines the crucial motivation of reductivism.

It is helpful to introduce this argument by addressing the basic question of why philosophers are troubled by the failure of reductive explanations, anyway. Arguably, one important motivation of reductivism is the explanatory success of modern science. Given well-known scientific success stories such as the explanation of life, one may assume that similar explanations should be also possible in the case of entities such as consciousness.[1] Furthermore, one may turn this analogy into an inductive argument: science has been successful in reductively explaining entities x_1-x_n (e.g. life, water, temperature…) through more fundamental entities y_1-y_n. Through inductive generalization, we are led to the assumption that we should also expect reductive explanations of currently unreduced entities including consciousness.

The inductive argument may be innocent if it is only supposed to show that we should not be too quick in ruling out the possibility of a reductive explanation of consciousness. In fact, I am sympathetic to the claim that the history of science should make us suspicious of armchair claims about the explanatory limits of the cognitive sciences. However, this claim does not lead to reductivism and is entirely compatible with a relaxed pluralist attitude that considers the scope of reductive explanations an open empirical question. An inductive argument for reductivism would have to show that we should expect *all* entities to be reductively explicable without any exceptions whatsoever. Unfortunately, it is very hard to see how examples of successful reductive explanations could possibly justify this general reductivist stance. First, one can object that a sober look at scientific practice indicates that reductive explanations are usually not a tenable goal and that we actually find a huge diversity of scientific explanations that include traditional forms of theory reductions and reductive explanations but also many forms of non-reductive explanation and integration. In fact, much of the contemporary literature in philosophy of neuroscience challenges traditional models of reductive explanations and insists on the dominance of more moderate forms of – e.g. mechanistic – explanation in scientific practice (cf. Gervais and Looren de Jong 2013).

Of course, a reductivist can respond that reductive explanations are still possible "in principle" and that mechanistic explanations will eventually enable us to achieve reductive explanations in a more narrow philosophical sense (e.g. Chalmers 2012, 306). However, these in principle-claims put the burden of proof on a reductivist who has to give us reasons to believe that everything is in principle reductively explicable. As Eronen (2011, 138) puts it: "An unrelenting reductionist might still

---

[1] Philosophers of mind often refer to the demise of vitalism and "the explanation of life" in rather questionable ways that do not do justice to historical vitalist positions (cf. Normandin and Wolfe 2013). For the sake of the argument, however, I will leave this piece of whig history unquestioned.

claim that there is 'in principle' derivability from lower to higher levels, meaning that given enough computational power and time, we could use the molecular level generalizations to explain anything the higher-level generalizations explain. However, how could we evaluate such 'in principle' claims, given that we do not have the time and the computational power, and we do not know what the 'completed' sciences will look like?"

A second problem with the inductive argument is that the well-known problems of reductive explanations can be interpreted as counterexamples to a reductivist induction. For example, I have argued that there are good reasons to reject reductive explanations of many biological kinds including species. Furthermore, one can make the same argument with traditional "placement problems" in metaphysics such as abstract objects, consciousness, indexicality, or normativity. Why shouldn't we consider these examples as the black swans that undermine any inductive attempt to justify reductivism?

Even if inductive considerations do not provide more than an "intuition pump" in the sense of Dennett (1988), many philosophers will insist that we still have good reasons to believe that reductivism is true. Arguably, the most important case for reductivism starts with metaphysics instead of inductive considerations from scientific practice: reductivism is necessary to make sense of our metaphysical commitment to a fundamentally physical reality. For example, my historical discussion in the introduction quoted Smart as claiming that "on this view there are, in a sense, no sensations. A man is a vast arrangement of physical particles but there are not over and above this, sensations or states of consciousness (1959, 142)." Given that we are fundamentally "vast arrangements of physical particles", it seems plausible and maybe even inevitable to assume that reductive explanation must be (at least in principle) available.

One way of justifying this assumption of a fundamentally physical reality is based on the general ideal of exactly one fundamental ontology. Only physics describes what exists in the most fundamental sense and a fundamental ontology will therefore be a purely physical ontology. It is not hard to see how this ideal of a fundamental physical ontology can motivate reductivism: if only physics describes what fundamentally or really exists, it seems almost trivial that all legitimate non-fundamental entities have to be somehow explicable in terms of the fundamental physical entities. It is this commitment to a fundamental physical ontology that seems to make reductivism in philosophy of mind a well-justified assumption. If it were not possible to physically explain phenomenal states, there would be no place for phenomenal entities in the fundamental physical world.

The idea of exactly one fundamental physical ontology may justify reductivism but is challenged by conceptual relativity. Conceptual relativity challenges the *very idea* of exactly one fundamental ontology no matter whether this ontology characterized in physicalist or non-physicalist terms. While this challenge clearly illustrates a tension between conceptual relativity and a common motivation of reductivism, reductivists can respond by proposing a substantive account of the ontological propriety of the physical that does not presuppose the ideal of exactly one fundamental ontology. Even if we do not assume that only physics describes what fundamentally exists, the physical may still be ontologically prior in a sense that is strong enough to justify reductivism. The goal of the following sections is to argue that such an account is not available. More specifically, I will argue that all available accounts of the ontological priority of the physical turn out to be (1) too weak to justify reductivism, (2) circular in already presupposing reductivism, or (3) incompatible with conceptual relativity.

This situation leads to the conclusion that conceptual relativists have no reason to endorse global reductivism: the only non-circular notion of ontological priority that is strong enough to motivate reductivism is incompatible with conceptual relativity. Furthermore, I argue that this "argument from ontological non-fundamentalism" leads to a non-reductivism that should be distinguished from stronger variants of anti-reductivism as it considers the scope of reductive explanations an open empirical question. The goal of the argument is not to show that certain entities cannot be reductively explained but rather that we have no reasons to assume that they must be reductively explained. Whether reductive explanations are available will be determined by the empirical sciences and neither of the possible outcomes are philosophically puzzling.

7.1   Notions of Ontological Priority

According to the argument from ontological non-fundamentalism, conceptual relativity undermines the crucial motivation of reductivism by challenging the assumption that only physics describes what fundamentally exists. Reductivists can respond to this challenge by proposing an account of the ontological priority of the physical that is compatible with conceptual relativity but strong enough to motivate reductivism. In order to be compatible with conceptual relativity, a notion ontological priority cannot take the following form:

(a) An ontology O1 is prior to an ontology O2 if only O1 describes what fundamentally exists.

According to (a), the physical is ontologically prior to everything else if only a physical ontology describes what fundamentally exists. Ontological priority of the physical in the sense of (a) would arguably be strong enough to motivate reductivism: if only physics describes what fundamentally exists and if we want to keep an entity in our ontology, we have to explain this entity in terms of physical entities. For example, if we want to keep consciousness in our ontology, we have to explain consciousness in terms of physical entities. At the same time, this ideal of exactly one fundamental physical ontology is challenged by conceptual relativity as the very point of conceptual relativity is the rejection of the idea of exactly one fundamental ontology. A notion of ontological priority that is compatible with conceptual relativity needs to provide an alternative account of the priority of physical ontologies.

One obviously unsatisfying option is a notion of ontological priority that is tied to reductive explanation:

(b) An ontology O1 is prior to an ontology O2 if the entities of O2 can be reductively explained through the entities of O1

According to (b), the physical level is ontologically prior if everything is reductively explainable in terms of physics. Although (b) provides a very intuitive account of priority, it cannot be the notion of ontological priority we are looking for in the present context. As we are looking for a notion of ontological priority that will justify reductivism, we cannot presuppose reductivism in our justification of the ontological priority of the physical.

Neither (a) nor (b) offer a satisfying explanation of the ontological priority of the physical that is compatible with conceptual relativity and non-circular in its justification of reductivism. This does not mean that every possible notion of the ontological priority of the physical is either incompatible with conceptual relativity or circular. Consider the following definition:

(c) An ontology O1 is prior to an ontology O2 if the objects of O2 are composed by the objects of O1.

Claims about composition in the sense of (c) are often an important part in reductivist frameworks. Furthermore, we can extend the claim from objects to properties. A property F is ontologically prior to a property G if the following conditions are met: F(a) & G(b) & b is composed of a. Unfortunately, there are important problems with any justification of ontological priority in the sense of (c). First, contemporary physics challenges the traditional picture of mereological composition according to which all objects can be understood as composites of a fundamental level.[2] Second, a notion of ontological priority in the sense of (c) would be clearly too weak to justify reductivism. Even if we assume a traditional picture of mereologically structured layers of reality, it is hard to see how composition could imply a notion of ontological priority that would justify reductivism. For example, many contemporary philosophers reject the idea that composing parts should be considered prior to the composed whole (e.g. Hüttemann 2003; Schaffer 2010).

The claim that all material objects are composed of physical objects is not the only comparably weak interpretation of the priority of the physical. One rather uncontroversial observation is that physical ontologies are more general than most other ontologies in the sense that they have a larger scope of application. For example, every biological organism can also be described as a physical object, while not every physical object can be described as a biological organism. The scope of application suggests another interpretation of ontological priority:

(d) An ontology O1 is prior to an ontology O2 if O1 has a larger scope of application than O2.

---

[2] This point is already a crucial motivation in Suppes' landmark paper "on the plurality of science" (1978). For a more recent and more detailed criticism of speculations about mereological composition from a fundamental level, see Ladyman and Ross (2007).

Given this definition of ontological priority, it seems plausible to argue that physical ontologies should be considered prior to all other ontologies. Of course, there are some obvious difficulties such as abstract objects, meaning, indexicality, or norms that may not be in the scope of physical ontologies but I am going to set these problems aside for the sake of the argument. Even if we assume that physics is prior to everything else in the sense of (d), we would still have to show that this notion of ontological priority justifies reductivism. It seems clear that this is not the case. The fact that O1 has a larger scope of application than O2 certainly does not imply that there must be a reductive explanation of O2 in terms of O1.

Consider a real-life example of ontologies with different scopes of application. The *International Classification of Diseases* (ICD) of the World Health Organization has a larger scope of application than the *Diagnostic and Statistical Manual of Mental Disorders* (DSM) of the American Psychiatric Association because the latter only includes mental disorders while former includes both mental disorders and other diseases. Even if it may be helpful in some contexts to say that the ICD is therefore "prior" to the DSM, this is obviously a very weak notion of priority that does not carry any implications about reductions or reductive explanations.

7.2 Supervenience-Based Formulations of Ontological Priority

So far, the discussion of different notions of ontological priority has not been of any help for reductivists. There are innocent interpretations of the claim that the physical is ontologically prior to everything else. At the same time, these innocent interpretations are not strong enough to justify reductivism. If we turn to more ambitious notions of ontological priority such as (a) and (b), we either have to presuppose the idea of exactly one fundamental ontology or reductivism. Therefore, none of these interpretations are helpful in justifying reductivism under the assumption of conceptual relativity. At the same time, (a)–(d) are by no means the only possible interpretations of ontological priority and I anticipate the objection that a more far-reaching notion might be based on the notion of supervenience. Let us consider the following proposal:

(e) An ontology O1 is prior to an ontology O2 if the entities of O2 supervene on the entities of O1.

As there are many different supervenience concepts, (e) is too vague to qualify as a satisfying interpretation of ontological priority. The weakest form of supervenience is based on generalized observations. Let us suppose we observe that every object that has the property F has also the property G, while not every object that has the property G has the property F. For example, we might observe that fluids with the same physical structure have the same macroscopic properties such as the same boiling point. At the same time, two fluids can have the same boiling point without having the same physical structure. In this sense, macroscopic properties such as a "boiling point of 100 °C" supervene on physical properties, while physical properties do not supervene on macroscopic properties such as "boiling point of 100 °C." As long as this observation is not backed by any modal claims, supervenience comes down to what might be called "*de facto* supervenience" (cf. McLaughlin 1995, 18). G supervenes on F if there is no change in G without a change in F. *De facto* supervenience suggests the following specification of (e):

(e') An ontology O1 is prior to an ontology O2 if the entities of O2 *de facto* supervene on the entities of O1.

There can be no doubt that (e') is not suited to justify a substantive notion of ontological priority that supports reductivism. Consider the following case of *de facto* supervenience: every living organism with an arthropod exoskeleton is an invertebrate, while not every invertebrate has an arthropod exoskeleton. In this sense, the property of not having a backbone (i.e. being an invertebrate) supervenes on the property of having an arthropod exoskeleton, while the property of having an arthropod exoskeleton does not supervene on the property of not having a backbone. However, it would be odd to claim that "having an arthropod exoskeleton" is ontologically prior to "not having a backbone." And even if we would find a way to make sense of this claim, this notion of ontological priority could not support reductivism. Obviously "not having a backbone" cannot be reductively explained by "having an arthropod exoskeleton." Therefore, *de facto* supervenience cannot provide the notion of ontological priority we are looking for.

I do not think anyone would challenge the claim that *de facto* supervenience is too weak to support reductivism. At the same time, the limits of de facto supervenience point toward a more promising formulation. In the case of the exoskeleton and backbone, the supervenience relation is unsatisfying because of its contingency. Even if every living organism with an arthropod exoskeleton is an invertebrate, we can easily imagine a different evolutionary history with organisms that have arthropod exoskeletons *and* backbones. In this case being an invertebrate would not supervene on having an arthropod exoskeleton.

Some supervenience relations are different from this case of contingent *de facto* supervenience. Recall the example of macroscopic properties such as "having a boiling point of 100 °C." Every fluid with a specific physical structure P has a boiling point of 100 °C, but not every fluid with a boiling point of 100 °C has the same physical structure P. This case differs from the biological example because it does not seem contingent that the macroscopic properties supervene on physical structure. In an important sense, the supervenience relation seems to be *necessary* and, therefore, suggests a different interpretation of ontological priority:

(e'') An ontology O1 is prior to an ontology O2 if the entities of O2 *necessarily* supervene on the entities of O1.

A formulation such as (e'') is still too vague because it does not specify *in what sense* the supervenience relation is necessary. One possible specification is known as "nomological supervenience." The entities described through O2 *necessarily* supervene on the entities described through O1, if there is a law that connects O2-properties with O1-properties.

(e''') An ontology O1 is prior to an ontology O2 if the entities of O2 *nomologically* supervene on the entities of O1.

Even if we accept that (e''') offers a legitimate interpretation of priority, it should be uncontroversial that this is a case of nomological and not ontological priority. Consider epiphenomenalism as an obvious illustration: according to epiphenomenalism, the physical is *nomologically* prior to the mental as physical states cause mental states, while mental states do not cause physical states. At the same time, epiphenomenalism denies the ontological priority of the physical. Physical and mental states are metaphysically distinct; neither of them can claim ontological priority in a sense that would be strong enough to justify reductivism. The case of epiphenomenalism illustrates why (e''') cannot provide the notion of priority that we are looking for. Nomological supervenience does not justify reductivism and does not imply a substantive account of the ontological priority of the physical.

While nomological supervenience is too weak to justify ontological priority claims, *metaphysical* supervenience seems to constitute a more promising option. Physicalism is often introduced as the idea that everything metaphysically supervenes on the physical. For example, mental properties supervene on physical properties with metaphysical necessity in the sense that mental differences without physical differences are metaphysically impossible. We can therefore formulate another supervenience based notion of fundamentality:

(e'''') An ontology O1 is prior to an ontology O2 if the entities of O2 *metaphysically* supervene on the entities of O1.

Unfortunately, there are good reasons to doubt that (e'''') will lead to a notion of ontological priority that is strong enough to justify reductivism. It is helpful to distinguish between two types of objections.[3] On the one hand, one can doubt that (e'''') will lead to any substantive notion of ontological priority whatsoever (cf. Daly 1997). On the other hand, one can grant that (e'''') leads to a substantive notion of ontological priority but argue that this notion is not strong enough to justify reductivism. This is strategy is endorsed by many physicalists who claim that everything metaphysically supervenes on the physical but deny that everything can be reductively explained in terms of the physical.

The question whether metaphysical supervenience implies ontological priority at all is nicely illustrated by Stoljar's example of a "necessitation dualist" who holds that "while psychological and physical properties are metaphysically distinct, they are nevertheless of laws of nature necessarily connected: that is, in all possible worlds, if various physical properties are instantiated then so too are various

---

[3] In the spirit of a strong "naturalism of scientific practice" one could also challenge the intelligibility of metaphysical supervenience because of its reliance on a dubious notion of metaphysically possible worlds. For the sake of the argument, however, I will set aside this general modal skepticism and grant metaphysicians that there is some stable notion of metaphysical supervenience.

psychological properties" (Stoljar 2010, 145). In other words: the necessitation dualist accepts that the mental metaphysically supervenes on the physical but rejects the ontological priority of the physical.

Proponents of supervenience-based accounts of physicalism can react to the alleged counterexample of necessitation dualism by denying its coherence. One reason for rejecting the coherence of necessitation dualism is "Hume's Dictum" according to which there are no metaphysically necessary connections between metaphysically distinct entities (for more careful formulations see Stoljar 2010 and Wilson 2010). For example, if mental properties metaphysically supervene on physical properties, Hume's Dictum implies that they are not metaphysically distinct. If Hume's Dictum is true, then the claim that everything metaphysically supervenes on the physical implies that dualism is wrong and that necessitation dualism is not even an option. Why, however, should a necessitation dualist accept Hume's Dictum? As both MacBride (2005, 125) and Wilson (2010, 596) have pointed out, there is a surprising shortage of arguments for Hume's Dictum given its crucial role as a premise in many arguments of contemporary metaphysicians.

One possible reply is that Hume's Dictum does not need much justification as it is analytic. Furthermore, it is also not hard to see how one can interpret Hume's Dictum as analytic by equating metaphysical distinctness with modal distinctness. Given this interpretation, Hume's Dictum simply states that there are no metaphysically necessary connections between modally distinct entities. While this interpretation may render Hume's Dictum analytic, necessitation dualists won't be impressed by this argument. As necessitation dualists acknowledge metaphysical supervenience, they are also committed to the claim that mental and physical entities are not *modally* distinct in the sense of this interpretation. However, necessitation dualists still claim that mental and physical entities are *metaphysically* distinct and that any claim about modal distinctness simply misses the point.

Furthermore, even if we accept Hume's Dictum, one can doubt that metaphysical supervenience justifies a substantive notion of ontological priority because it fails to exclude some nondualist competitors of physicalism. If we accept Hume's Dictum and the claim that everything metaphysically supervenes on the physical, we establish the conclusion that there are no entities that are metaphysically distinct from physical entities. It should be uncontroversial that this monist conclusion is sufficient to exclude any substantive variants of dualism and metaphysical pluralism. However, monism comes in three flavors: physicalism, neutral monism, and idealism. While all three types of monism can agree on the rejection of metaphysical distinctions between physical and mental entities, they disagree on issues of ontological priority. Physicalists insist on the priority of the physical, idealists insist on the priority of the mental, and neutral monists reject both priority claims. The conjunction of metaphysical supervenience and Hume's Dictum therefore seems to be at best a general formulation of monism but not of physicalism.

To stress this point, we can introduce a "necessitation monist" in analogy to Stoljar's necessitation dualist who accepts that mental and physical properties are necessarily connected and that they are not metaphysically distinct. However, the necessitation monist rejects the ontological priority of the physical in the same sense as she rejects idealism by refusing to accept the ontological priority of the mental. Furthermore, my discussion of "psychophysical parallelism" in chapter 2 provides a historical example of "necessitation monism". Recall that Moritz Schlick, one of the most prominent proponents of this position, presented "psychophysical parallelism [as] a harmless parallelism of two differently generated concepts" (Schlick 1927). Schlick insists that his psychophysical parallelism implies that mental and physical states are *not* metaphysically distinct as it is only "an epistemological parallelism between a psychological conceptual system on the one hand and a physical conceptual system on the other. The 'physical world' *is* just the world that is designated by means of the system of quantitative concepts of the natural sciences." (1918/1974, 301). Whereas dualists assume a metaphysical gap, Schlick argues for a conceptual gap. "There is only *one* reality," (1918/1974, 244) but this reality can be described in terms of different conceptual systems. Schlicks parallelism is arguably compatible with metaphysical supervenience and still rejects ontological priority claims and therefore illustrates that (e"") will not lead to the desired notion of the priority of the physical.

So far, I have presented "necessitation dualism" and "necessitation monism" as objections against attempts to formulate the ontological priority of the physical in terms of metaphysical supervenience. However, there is a further problem. Even if we grant that metaphysical supervenience implies ontological priority, one can claim that this notion of "ontological priority" will not be strong enough to justify reductivism. In fact, metaphysical supervenience is especially popular among non-reductive physicalists because it allows a separation of metaphysical claims about ontological priority and epistemological claims about reductive explanations. However, this separation of metaphysical and epistemological claims makes metaphysical supervenience unattractive for a motivation of reductivism.

If (e')-(e"") are too weak to motivate reductivism, one may attempt to formulate an even stronger interpretation of supervenience relations. In *The Conscious Mind*, Chalmers offers such an interpretation. According to Chalmers' notion of logical supervenience, "B-properties supervene *logically* on A-properties if no two *logically possible* situations are identical with respect to their A-properties but distinct with respect to their B-properties."[4] The case of macroscopic properties such as "having a boiling point of 100 °C" can illustrate Chalmers' notion of logical supervenience. I already mentioned that the supervenience relationship between the boiling point of a fluid and its physical structure seems to be a necessary relationship. Philosophers often assume that this is not because the physical structure of a fluid is nomologically connected with its macroscopic properties, but because the macroscopic properties *are derivable* from certain physical properties. Given a comprehensive physical theory and an appropriate conceptual analysis, we can deduce that water has a boiling point of 100 °C. In this sense, no two *logically possible*

---

[4] See Chalmers (1996, 42–51). A similar notion is Terry Horgan's "superduperveniece," on which see Horgan (1993).

situations are identical with respect to their physical properties but distinct with respect to their boiling point. Logical supervenience implies another interpretation of priority:

(e''''') An ontology O1 is prior to an ontology O2 if the entities of O2 *logically* supervene on the entities of O1.

Does (e''''') provide a notion of ontological priority that justifies reductivism? While (e''''') implies a strong notion of ontological priority, it is as circular as (b) when used as a justification of reductivism. Logical supervenience presupposes logical derivability of reduced phenomena. Anyone how has doubts about reductive explanation will have the same doubt about logical supervenience. (e''''') is therefore obviously not suited as a justification of reductivism. Indeed, if we had reasons to believe that everything logically supervenes on the physical, we would trivially also have reasons to believe that everything is reductively explicable in terms of the physical. At the same time, doubts about reductivism extend to doubts about logically supervenience and (e''''') does therefore not help to answer the question why we should assume reductivism in the first place.

To sum up, I have discussed a variety of notions of priority. (a) a notion of priority that directly contradicts conceptual relativity by claiming that O1 is more fundamental than O2 if only the entities of O1 fundamentally exist. (b) a formulation of priority in terms of reductive explanation. (c) / (d) less ambitious notions of priority in terms of composition or scope of application. (e')-(e''''') a large range of notions of priority in terms of supervenience. However, none of these proposals meet the following three conditions:

(i) it justifies reductivism in the sense of the assumption that everything has to be physically explained (violated by c, d, e', e'', e''', e'''');
(ii) it is not circular in the sense that it doesn't presuppose reductivism (violated by b and e''''');
(iii) and it is compatible with conceptual relativity (violated by a).

This situation leaves reductivism in a highly uncomfortable position. Arguably, reductivism is crucially motivated by the ontological priority of the physical. We should expect reductive explanations *because* the physical is ontologically prior to everything else. This motivation of reductivism requires a notion of the ontological priority of the physical that does not already presuppose reductivism. While there are a variety of notions of the priority of the physical that do not presuppose reductivism (c, d, e', e'', e''', e''''), I have argued that none of them are strong enough to justify reductivism. The only non-circular notion of ontological priority that is strong enough to justify reductivism is based on the ideal of exactly one fundamental ontology and the claim that only physics describes what exists in the most fundamental sense. The motivation of reductivism therefore turns out to be incompatible with conceptual relativity.

7.3   Reductivism, Non-reductivism, and Anti-reductivism

None of the formulations (a)–(e) offer a non-circular account of ontological priority that is compatible with conceptual relativity and still strong enough to justify reductivism. Therefore, proponents of conceptual relativity should reject the ontological priority claims that crucially motivate reductivism. Given the assumption of conceptual relativity, there is simply no reason to believe that everything must be explicable in terms of the physical. It is important to distinguish this non-reductivist conclusion from more familiar anti-reductivist claims that are common in contemporary philosophy of mind. Anti-reductivism and non-reductivism come with different attitudes regarding reductive explanation (and reduction) that can be expressed through the following templates:

Anti-reductivism: "x cannot be reductively explained."
Non-reductivism: "it is an open empirical question whether x can be reductively explained."

While the *argument from horizontal pluralism* aims at an anti-reductivism regarding certain entities such as biological patterns, the goal of the *argument from ontological non-fundamentalism* is to justify a more general non-reductivism regarding controversial entities such as consciousness. The development of the cognitive sciences may or may not lead to a reductive explanation of consciousness and we have no reason to believe that things must turn out one way or another.

In the introduction, I presented this non-reductivist attitude as variety of naturalism that takes the diverse ontologies in scientific practice as a starting point for philosophical reasoning. Philosophers are not in the epistemic position to step behind our scientific ontologies by telling us what entities fundamentally exist. If we want to know what scientific entities exist, we need to look at the ontologies that are used in successful science. The same naturalistic reasoning applies to issues of reduction and reductive explanation: If we want to understand the prospects and limits of reduction and reductive explanation, we need to start with a sober look at scientific practice and avoid the temptations of presupposing a "grand metaphysical story" that imposes a uniform model of scientific explanation and of relations between scientific ontologies.

The main claim of non-reductivism is that the scope of reductions and reductive explanations is an open empirical question. Of course, this does not mean that philosophers have nothing positive to contribute to debates about reductions, reductive explanations, and scientific explanations in general. These contributions, however, have to be grounded in detailed engagement with scientific practice and will most likely lead to a complex picture that includes large variety of reductive unifications, non-reductive forms of integration, and stronger forms of disunity.[5]

---

[5]Abney et al.'s (2014) discussion of joint perceptual decision-making provides a helpful example of explanatory pluralism and of the crucial importance of non-reductive forms of integration in cognitive science. Their discussion focuses on three approaches to decision making ("behavioral/decision-making", "linguistic/confidence", and "physical/acoustic energy") and the relations

Contemporary philosophy of neuroscience provides helpful examples of more complex models of scientific explanation that are based on engagement with detailed case studies and not on a presupposed metaphysical picture. The most prominent examples are current debates about mechanistic explanation (e.g. Brigandt 2013, Gervais and Looren de Jong 2013; Piccinini and Craver 2011; Woodward 2008) that are often presented in sharp contrast with traditional philosophical accounts of theory reduction and reductive explanation.

Mechanistic philosophers of neuroscience often stress that their accounts aim at analyzing what neuroscientists *actually* do. Furthermore, these explanations in scientific practice usually do not neatly fit the reducible/irreducible dichotomy. Mechanistic explanations share some features with traditional theory reductions and reductive explanations. Most importantly, they appeal to lower-level entities and consider the component parts of a mechanism a crucial aspect in its explanation. At the same time, they do not consider the components and their properties sufficient for the explanation of a mechanism – knowledge about mechanisms cannot be simply derived from knowledge about its components. Craver stresses this point when contrasting mechanistic explanations with reductions: "The central idea is that neuroscience is unified not by the reduction of all phenomena to a fundamental level, but rather by using results from different fields to constrain a multilevel mechanistic explanation" (2007, 231).

Many mechanistic philosophers of neuroscience combine this analysis with an explanatory pluralism that acknowledges the diversity of scientific explanations that often require multiple explanatory levels. Craver even states that his "view is consistent with, although it is a much more precise specification of, Dupre's pluralistic epistemology" (2007, 268). At the same time, Craver avoids any commitment to ontological pluralism but "leave[s] open the question of whether metaphysical arguments can be mustered to support a principled ontological fundamentalism" (2007, 13).

The avoidance of ontological issues is well justified in the context of a philosophy of neuroscience that aims at analyzing neuroscientific explanations and self-identifies as a part of philosophy of science instead of philosophy of mind. Furthermore, it reflects debates about reduction in other areas of philosophy of science that are not concerned with issues of ontological unification (cf. 6.3). At the same time, an ontological agnosticism allows reductivist philosophers to argue for

---

between them. Although the approaches cannot be reduced to each other, there are many substantive forms of theory integration. The different approaches do not only provide substantive and unique information about joint decision-making but also illuminate each other's research results. As Abney et al. put it: "If none of these three approaches informed each other, various research opportunities crossing scales and mixing methods would be lost. For example, can people with asymmetric perceptual capabilities effectively overcome their difference by communicating about common environmental constraints (Approach 1) and what do the language properties (Approach 2) look like when they successfully coordinate?" (2014, 9) The example of joint perceptual decision-making is therefore a helpful reminder that scientific practice is often concerned with meaningful forms of integration and explanation that have little use for the opposition between reductionism and anti-reductionism.

the compatibility of mechanistic explanations in philosophy of neuroscience and more ambitious forms of reductive explanations as ideals in debates about the metaphysics of the mind. More specifically, philosophers of mind can argue that the ontological commitment to exactly of fundamental (e.g. physical) ontology justifies the stronger assumption that everything must be – not in actual scientific practice but *in principle* – explicable in terms of a fundamental physical level. Even if contemporary scientific practice is dominated by more moderate forms of scientific explanation, a fully developed neuroscience would have to provide substantive reductive explanations that satisfy ontological fundamentalism.

My discussion of conceptual pluralism suggests that we should not presuppose ontological fundamentalism but extend the empirically grounded methodology of contemporary philosophy of neuroscience from issues of explanation to issues of ontology. In the same way as discussions of scientific explanation should start with an analysis of scientific practice instead of a presupposed philosophical model of reduction, a discussion of ontological unification should also be empirically grounded.

References

Abney, Drew, Rick Dale, Jeff Yoshimi, Chris Kello, Kristian Tylén, and Riccardo Fusaroli. 2014. Joint Perceptual Decision-Making: A Case Study in Explanatory Pluralism. *Theoretical and Philosophical Psychology* 5 (330): 1–19.

Brigandt, Ingo. 2013. Systems Biology and the Integration of Mechanistic Explanation and Mathematical Explanation. *Studies in History and Philosophy of Science Part C: Studies in History and Philosophy of Biological and Biomedical Sciences* 44 (4): 477–492.

Chalmers, David. 1996. *The Conscious Mind: In Search of a Fundamental Theory*. Oxford: Oxford University Press.

Chalmers, David. 2012. *Constructing the World*. Oxford: Oxford University Press.

Craver, Carl F. 2007. *Explaining the Brain*. Oxford: Oxford University Press.

Daly, Chris. 1997. Pluralist metaphysics. *Philosophical studies* 87 (2): 185–206.

Dennett, Daniel C. 1988. Quining Qualia. In *Consciousness in Modern Science,* eds. Anthony J. Marcel and Eduard Bisiach, 42–77. Oxford: Oxford University Press.

Eronen, Markus I. 2011. *Reduction in Philosophy of Mind: A Pluralistic Account*. Berlin: Walter de Gruyter.

Gervais, Raoul and Looren De Jong, Huib 2013. The Status of Functional Explanation in Psychology: Reduction and Mechanistic Explanation. *Theory and Psychology* 23 (2): 145–163.

Horgan, Terence. 1993. From Supervenience to Superdupervenience: Meeting the Demands of a Material World. *Mind* 102 (408): 555–86.

Hüttemann, Andreas. 2003. *What's Wrong with Microphysicalism?* New York: Routledge.

Ladyman, James, and Don Ross. 2007. *Every Thing Must Go: Metaphysics Naturalized*. Oxford: Oxford University Press.

MacBride, Fraser. 2005. Lewis's Animadversions on the Truthmaker Principle. In *Truthmakers: The Contemporary Debate*, eds. Helen Beebee and Julian Dodd, 117–40. Oxford: Oxford University Press.

McLaughlin, Brian. 1995. Varieties of Supervenience. In *Supervenience: New Essays*, eds. Elias E Savellos and Ümit D Yalçin, 16–59. Cambridge: Cambridge University Press.

Normandin, Sebastien, and Charles T. Wolfe. 2013. *Vitalism and the scientific image in post-enlightenment life science, 1800-2010*. Berlin: Springer.

Piccinini, Gualtiero, and Carl Craver. 2011. Integrating Psychology and Neuroscience: Functional Analyses as Mechanism Sketches. *Synthese* 183 (3): 283–311.

Schaffer, Jonathan. 2010. Monism: The Priority of the Whole. *Philosophical Review* 119 (1): 31–76.

Schlick, Moritz. 1918/1974. *General Theory of Knowledge*. Vienna: Springer.

Schlick, Moritz. Letter to Ernst Cassirer||, 1927. Inv. No. 94. Schlick-Paper

Smart, John JC. 1959. Sensations and Brain Processes. *The Philosophical Review* 68 (2): 141–56.

Stoljar, Daniel. 2010. *Physicalism*. New York: Routledge.

Suppes, Patrick. 1978. The Plurality of Science. *PSA: Proceedings of the Biennial Meeting of the Philosophy of Science Association*, 3–16.

Wilson, Jessica. 2010. What Is Hume's Dictum, and Why Believe It? *Philosophy and Phenomenological Research* 80 (3): 595–637.

Woodward, James. 2008. Mental Causation and Neural Mechanisms. In *Being Reduced*, eds. Jakob Hohwy and Jesper Kallestrup, 52–74. Oxford: Oxford University Press.

Part IV
Beyond the Mind-Body Problem

Chapter 8
Consciousness

In the introduction, I sketched the strategy of a pluralist theory of the mind along the following lines:

1. Scientists describe reality in terms of different but equally fundamental ontologies.
2. Given a plurality of equally fundamental ontologies, there is no reason to consider limits of reduction and reductive explanations philosophically puzzling.
3. If there is no reason to consider limits of reduction and reductive explanations philosophically puzzling, then there is no reason to worry about explanatory gaps in philosophy of mind and common formulations of the mind-body problem are flawed.

∴ There is no reason to worry about explanatory gaps in philosophy of mind and common formulations of the mind-body problem are flawed.

The first premise has been defended in the chapters on conceptual relativity and the second premise has been backed by the argument from horizontal pluralism and the argument from ontological non-fundamentalism. The aim of this last part of the book is to defend the claim that the proposed conceptual pluralism does indeed lead to the conclusion that the most common formulations of the mind-body problem are flawed.

I will consider three challenges to the claim that conceptual pluralism undermines the philosophical obsession with explanatory gaps. First, I will discuss some well-known formulations of the mind-body problem and address the question whether conceptual pluralism offers a plausible answer to the so-called "hard problem of consciousness" (Chap. 8). Second, I will consider the objection that conceptual pluralism is an unstable position that collapses back into physicalism or dualism (Chap. 9). Finally, I will briefly discuss the objection that my discussion of conceptual pluralism has missed an essential aspect of the mind-body problem: mental causation (Chap. 10).

The responses to all three challenges crucially depend on my presentation of conceptual relativity. Following Putnam, I have argued that a plurality of conceptual

© Springer International Publishing Switzerland 2015
D. Ludwig, *A Pluralist Theory of the Mind*, European Studies
in Philosophy of Science 2, DOI 10.1007/978-3-319-22738-2_8

frameworks implies a plurality of equally legitimate ontologies. For example, biologists rely on different species ontologies, cognitive scientists on different accounts of cognition, and psychologists on different intelligence ontologies. The same point can be made with traditional philosophical examples such as Putnam's universe of three elementary particles in an empty space. Nihilists and universalists propose different ontologies that provide different accounts of what it means for an object to exist.

Conceptual pluralism in this sense has to be distinguished from both a merely epistemological pluralism and a strong metaphysical pluralism. Contrary to merely epistemological pluralism, conceptual pluralism insists that conceptual relativity does not only imply epistemic but also ontological plurality. At the same time, this ontological pluralism has to be distinguished from metaphysically ambitious forms of ontological pluralism that reject conceptual relativity and aim at one fundamental pluralist ontology such as Karl Popper's theory of "three worlds" (1978). A metaphysically ambitious ontological pluralism can be seen as an extension of dualism: the fundamental picture of the world requires physical and non-physical mental entities but it also requires further non-physical and non-mental entities. While such a metaphysically ambitious pluralism will often run into the same problems as traditional forms of dualism, conceptual pluralism provides a metaphysically shallow interpretation of ontologies that alters the dialectical situation. In this section, I will argue that conceptual pluralism challenges traditional problems of consciousness and mental causation by allowing us to understand biological and psychological ontologies as different and equally fundamental conceptualizations of reality that do not imply any substantive form of metaphysical distinctness. Furthermore, I will try to show that such a conceptual pluralism constitutes a genuine alternative to both physicalism and dualism.

8.1 Revisiting the Hard Problem of Consciousness

Let us return to the basic question of why we should accept the existence of a mind-body problem in the first place. It has become a truism in philosophy of mind that the core of the mind-body problem is constituted by the "hard problem of consciousness." that comes in large variety of formulations. For example: "We do not see how to explain a state of consciousness in terms of its neurological basis. This is the Hard Problem of Consciousness." (Block 2002, 397). "How, in a basically material universe, are we to understand even the bare existence of consciousness?" (Shear 1999, 5). "It is widely agreed that experience arises from a physical basis, but we have no good explanation of why and how it arises." (Chalmers 1995, 7). These different formulations share the basic idea that it is not possible to explain consciousness in terms of a neurological/material/physical basis. But how is this a problem? Obviously, we need the additional premise, that it *should* be possible to explain consciousness in terms of a neurological/material/physical basis. But why should we accept this additional premise?

The most important reason why philosophers assume a reductive explanation should be possible is the ontological priority of the physical in the sense of the last chapter. If we consider only the physical ontologically fundamental, all non-physical entities must be in some sense derivative. If there is no reductive explanation of consciousness, however, consciousness also turns out to be fundamental and the ontological priority of the physical breaks down.

Given conceptual pluralism, this argument is everything but convincing and limits of reductive explanation hardly come as a surprise. Given the arguments from horizontal pluralism and ontological non-fundamentalism, there is no reason to assume that every entity *has to be* reductively explained in terms of a fundamental physical ontology. Instead, a "naturalism of scientific practice" starts with the diverse ontologies that we find in scientific practice and the equally diverse relations between them. Sometimes, these relations fit traditional or reformed models of reduction and reductive explanation. Sometimes, they don't. One way or another, we do not encounter a deep philosophical puzzle unless we presuppose a metaphysical picture that forces us to adopt a global reductivist strategy.

If there is no reason to adopt such a global reductivism, there is also no need to worry about limits of reductive explanation in the case of consciousness. At the same time, limits of reductive explanation do not imply that mental and physical concepts refer to metaphysically distinct entities in the sense of dualism or of a metaphysically ambitious pluralism. Instead, conceptual pluralism suggests that we often describe the same reality in terms of different but equally fundamental ontologies. Again, consider the example of Putnam's universe or case studies from the empirical sciences. Nihilists and universalists postulate the existence of different objects but do not refer to metaphysically distinct realms of reality in any substantive sense. Proponents of different species concepts postulate the existence of different kinds but do not refer to metaphysically distinct realms of reality in any substantive sense. Instead, they simply describe the same reality (Putnam's toy universe, the biological realm) in terms of different ontologies.

The rejection of (sufficiently strong variants of – cf. Sect. 6.3) reductionism and metaphysical distinctness raises the question whether my pluralist proposal leads to emergentism and an emergentist theory of the mind. For example, Sartenaer (2013) describes emergentism as a "'third way' between radical monism and pluralism" and furthermore defines the goal of emergentism as avoiding both "metaphysical dichotomy [...] and pure identity" (365). The similarities between my presentation of pluralism and Sartenaer's discussion of emergentism are striking and become even clearer in the light of his taxonomy of three different kinds of emergentism. Sartenaer distinguishes between theoretical emergence (rejection of reductionism), explanatory emergence (rejection of reductivism), and causal emergence (rejection of global causal reduction). My discussion of reductionism (Sects. 6.1, 6.2, 6.3, and 6.4), reductivism (Sects. 6.5, 6.6, and 6.7), and mental causation (Chap. 10) suggests that my pluralism can actually be interpreted as endorsing all three types of emergentism.

Although I do not have any general objections against this interpretation, I still avoid the label "emergentism" for pragmatic reasons. Despite the mentioned similarities, "emergentism" is also often associated with more ambitious metaphysical positions that postulate emergent (e.g. vital or phenomenal) properties that "arise" from a more fundamental physical level (cf. Sartenaer forthcoming). Conceptual pluralism provides at best a metaphysically shallow interpretation of this idea by assuming that the addition of new properties is a rather common and unproblematic consequence of our diverse conceptual resources. New research interests lead to new ways of describing reality and therefore to new ontologies that are not always reducible to other already existing ontologies. However, this does not indicate that phenomenal or other emergent properties "arise" from the physical in any metaphysically ambitious and unique sense.

This difference becomes especially obvious in debates about causal emergence. My rejection of causal reduction is simply an extension of conceptual relativity from debates about ontology to causality. In the same way as there is not only one correct way of talking about existence, there is not only one correct way of talking about causes or "causal powers". Conceptual pluralism therefore accepts overdetermination as an entirely harmless form of linguistic overdetermination (cf. Chap. 10). Many contemporary emergentists clearly want more than that and require that emergent properties come with novel causal powers that are incompatible with the causal closure of the physical (e.g. O'Connor and Wong 2005; Wong 2010, cf. Macdonald and Mcdonald 2010).[1] To sum up, conceptual pluralism may be best understood as suggesting a *metaphysically shallow* version of emergentism that considers emergent properties to be a common consequence of limits of ontological unification in scientific practice.

8.2   The Common Puzzles

Many philosophers of mind will remain unconvinced by my arguments and object that a relaxed pluralism misses the core of explanatory gap arguments and of the hard problem of consciousness. Even if conceptual pluralists accept that there is an explanatory gap, they insist that limits of reductive explanation are only troubling under the assumption of global reductivism. But does this response not miss the very point of the argument? Everyone (including conceptual pluralists) should accept that different domains are related. For example, phenomenal states are correlated with certain brain states, and it seems perfectly reasonable to ask *why* these psychophysical correlations exist.

---

[1] This problem also extends to Sartenaer's discussion of "causal emergence". One the one hand, he defines this type of emergence as the antithesis of causal reduction and I am happy to endorse the idea that we do not have to understand mental causation in terms of a more fundamental account of physical causation. On the other hand, Sartenaer takes the thesis to be synonymous with the claim that "the whole exhibits genuinely new causal powers". I'm not sure that I understand how this formulation can be compatible with Sartenaer's rejection of "metaphysical distinctness" and it seems to me that we should take a deflationist position that rejects the idea of one fundamental way of talking about "causal powers".

This objection, however, misunderstands the dialectical situation. Remember that conceptual pluralism is not a version of dualism or strong metaphysical pluralism. According to conceptual pluralism, phenomenal and physical concepts do not refer to entities that are distinct in a metaphysically substantive sense (cf. Sect. 9.4). Given this assumption, there is a trivial explanation for psychophysical correlations: phenomenal states are correlated with certain physical states *because* phenomenal and physical concepts do not refer to metaphysically distinct aspects of reality. If we have two distinct perspectives instead of two distinct aspects of reality, then correlations are everything but mysterious.

Recall, once again, conceptual relativity in the case of mereological nihilism and universalism. Whenever a nihilist counts three objects, a universalist counts seven objects. Whenever a nihilist claims that two objects exist, a universalist claims that three objects exist. Is this correlation mysterious? Should we say that it is left unexplained *why* the number of the nihilist's objects is correlated with the number of the universalist's objects? Of course not. There is no mystery because the nihilist's concepts and the universalist's concepts do not refer to distinct parts of reality in any metaphysically ambitious sense. This stance becomes clearer when it is applied to a number of well-known arguments and beaten paths of philosophy of mind.

Bats Consider Thomas Nagel's question "What is it like to be a bat?" According to Nagel, we do not know the answer to this question and every attempt to answer it seems hopeless. Nagel argues that our limited understanding of bats rests on the fact that we cannot explain subjective experience from a third-person perspective. Hence, we cannot explain the subjective experience of bats in terms of biology or physics. In addition, Nagel suggests that it is mysterious that we cannot explain the subjective experience of bats in terms of biology or physics. We can distinguish between two claims:

(a) We cannot explain the subjective experience of bats in terms of biology or physics
(b) It is mysterious that we cannot explain the subjective experience of bats in terms of biology or physics

A conceptual pluralist who adopts a naturalism of scientific practice may consider (a) at least a credible working hypothesis.[2] Given that phenomenal and biological ontologies are shaped by very different pragmatic contexts and evidence, it is indeed not surprising if phenomenal entities are not explicable in terms of biological or physical entities (cf. Sects. 8.4 and 8.5). However, a conceptual pluralist will challenge (b) and ask why this explanatory limit should be mysterious. Nagel's answer is based on the assumption that a "physical theory of mind *must* account

---

[2] As pointed out by an anonymous reviewer, this clearly depends on the involved notion of "explanation". If we adopt the dominant notions of "reductive explanation" in philosophy of mind in the tradition of Chalmers, Kim, and Levine (cf. Sects. 6.5, 6.6, and 6.7), (a) seems to be credible working hypothesis. However, this is clearly not the only notion of "explanation" and certainly not the most relevant notion of explanation in scientific practice, either (cf. Faye 2014, Chaps. 5, 6, 7, and 8 for a helpful overview).

for the subjective character of experience." (Nagel 1974, 445, emphasis added)
"For if the facts of experience — facts about what it is like for the experiencing
organism — are accessible only from one point of view, then it is a mystery how the
true character of experiences could be revealed in the physical operation of that
organism" (Nagel 1974, 442).

For a conceptual pluralist, however, there is no reason to believe that "the true
character of experiences" must be "revealed in the physical operation of that organism." If we do not presuppose the ontological priority of the physical operations,
there is also no need to suppose that the "true character of experience" has to be
found on a physical level. The claim that the "true character" is the "physical character" is nothing but the idea of the ontological priority of the physical. Hence, a
conceptual pluralist can accept the premise that we cannot explain the subjective
experience of bats in terms of biology or physics and still reject the idea that there
is anything mysterious about that.

Mary Another famous argument in the philosophy of mind that it is useful to discuss here is Frank Jackson's thought experiment about Mary:

> Mary is a brilliant scientist who is, for whatever reason, forced to investigate the world from
> a black and white room via a black and white television monitor. She specializes in the
> neurophysiology of vision and acquires, let us suppose, all the physical information there is
> to obtain about what goes on when we see ripe tomatoes, or the sky, and use terms like
> 'red', 'blue', and so on. [...]
> What will happen when Mary is released from her black and white room or is given a
> color television monitor? Will she learn anything or not? It seems just obvious that she will
> learn something about the world and our visual experience of it. But then it is inescapable
> that her previous knowledge was incomplete. But she had *all* the physical information. *Ergo*
> there is more to have than that, and physicalism is false. (Jackson 1982, 130).

If we read Jackson's argument as a general illustration of the mind-body problem,
we can again distinguish two claims:

(c) Mary has all the physical information and still learns something new about
    colors
(d) It is mysterious that Mary has all the physical information and still learns some-
    thing new about colors

Given that the argument from ontological non-fundamentalism suggests that the
scope of reductive explanations is an open empirical question, conceptual pluralists
do not necessarily have to accept (c). For example, Dennett (1993, 401) quite con-
vincingly argues that we have no idea what "complete physical information" would
imply and that there is no reason to trust our intuitions about the thought experiment. Given that we have no real epistemic grip on the Mary scenario, conceptual
pluralists may propose an agnostic attitude that assumes that it is an open question
what Mary would and would not know. Even if we assume that (c) is true for the
sake of the argument, however, conceptual pluralists will reject (d). Of course, the
problems with (d) are analogous to the problems with Nagel's assumption (b).
Conceptual pluralism rejects that we *must* be able to derive our non-physical

ontologies from a fundamental physical ontology and therefore also reject that there is anything mysterious about the Mary scenario.

Zombies   Let us finally consider the argument from the conceivability of philosophical zombies. A philosophical zombie is an organism that acts like a normal human being, but does not have any conscious experience. It is obvious that the existence of such zombies would be both shocking and puzzling, but why should the mere *conceivability* of philosophical zombies be mysterious? Many strange things are conceivable: a chocolate bar bigger than the Eiffel Tower, Willard van Orman Quine being the beauty queen of Albania 2010, philosophical zombies, and so on. But what are we to conclude from this? Again, we need two basic claims:

(e) Philosophical zombies are conceivable
(f) It is mysterious that philosophical zombies are conceivable

Proponents of (f) usually rely on arguments concerning the relation between conceivability and possibility (e.g. Chalmers 2010, Chap. 6). I will not go into detail on these arguments here as my treatment of them is predictable by now. The arguments of proponents of conceivability rely on Kripke-style thought experiments to show that many situations turn out to be not conceivable even if they seemed so at first glance. For example, if A is reducible to B, we can deduce the existence of A from B, and in this sense ~A&B is not conceivable. Consider the well-known example of water and H_2O. If we can reduce water to H_2O, we can deduce the existence of water from the existence of H_2O. In this sense, H_2O without water ("zombie water") is not conceivable, while zombies remain conceivable.[3]

However, we have already seen that a conceptual pluralist will not be puzzled by this kind of conceivability. First, the *argument from horizontal pluralism* suggests that there are countless cases of conceivability without possibility. For example, most biological entities refer to patterns that are not explicable in terms of a fundamental physical ontology and we often cannot deduce the existence of biological patterns from the existence of physical entities without endorsing a unacceptable eliminativism regarding biological entities (cf. Sect. 6.7). In this sense, biological patterns provide an example of ~A&B being conceivable but not possible.

Second, the *argument from ontological non-fundamentalism* suggests that we also lack positive evidence that conceivability always implies possibility. At best, examples such as Water and H_2O show that scientific practice also contains examples of reduction that undermine not only possibility but also conceivability. However, these examples provide a very shaky induction base for the claim that we will *always* find a link from conceivability to possibility.

---

[3] Not everyone accepts this presentation and Block and Stalnaker (1999) provide a highly influential critique of the standard models of reduction and reductive explanation of water (cf. Polger 2008; Bogardus 2013). However, I will grant reductivists these examples at least for the sake of the argument.

8.3   The Uniqueness of Phenomenal Concepts

Chalmers' classification of responses to the conceivability argument distinguishes between a type-a materialism that rejects the conceivability of philosophical zombies and a type-b materialism that rejects the inference from conceivability to possibility (Chalmers 2010, cf. Levin 2008; Webster 2006). Typically, non-reductive physicalists are type-b materialists in the sense of Chalmers and reject the possibility of philosophical zombies while accepting their conceivability. Non-reductive physicalists therefore usually share with conceptual pluralists some form of commitment to the epistemic autonomy of the phenomenal perspective. However, this epistemic pluralism of non-reductive physicalism is combined with a commitment to the ontological priority of the physical while I have introduced conceptual pluralism as a more radical position that rejects the very idea of exactly one fundamental ontology.

While I will argue that any satisfying non-reductivism requires the rejection of ontological fundamentalism, there are some important similarities between my proposal and many current forms of type-b materialism and non-reductive physicalism.[4] One convenient starting point for a discussion of these similarities is the so-called "phenomenal concept strategy" (PCS) that has emerged as one of the most fashionable frameworks for non-reductive physicalism. Although there is a huge variety of phenomenal concept strategies, they share the basic assumption that the explanatory gap is not due to differences between phenomenal and physical *states* but differences between phenomenal and physical *concepts*. There is no metaphysical gap but a fundamental conceptual gap which is why even physicalists can accept the irreducibility of phenomenal consciousness.

It is not difficult to see the similarities between PCS and my proposal of conceptual pluralism: both strategies accept the existence of an explanatory gap but argue that this gap can be understood in terms of different ways of conceptualizing reality. While conceptual pluralism can be understood as variety of PCS (cf. Ludwig 2013), I also argue that PCS can only be successful if it gives up the idea of exactly one fundamental ontology. PCS therefore does not lead to a "New Wave Materialism" (cf. Horgan and Tienson 2001; McLaughlin 2001) but rather to a "New Wave Pluralism" that should be distinguished from both dualism and physicalism.

Still, it will be instructive to have a closer look at common physicalist presentations of PCS and their accounts of differences between phenomenal and physical concepts. Thus far, my objections against the common puzzles of consciousness have been based on a general rejection of reductivism. I have argued that we should start with the diversity of meaningful explanations that we find in scientific practice and that we do not have to worry about the limits of reductive explanation. While I

---

[4] In fact, some outspoken epistemic pluralists like Eronen (2011) and Hüttemann (2003) also self-identify as physicalists. It is important to note that debates about the compatibility of pluralism and physicalism crucially depend on the involved notion of "physicalism". While I presuppose in the current discussion that physicalism requires a substantive account of ontological priority, more liberal definitions of "physicalism" are clearly possible (see also Sect. 9.2).

have therefore treated the limits of reductive explanation an open empirical question, I have not provided any argument for or against reductive explanations of our phenomenal perspective. Instead, the argument from horizontal pluralism has been concerned with non-phenomenal entities such as species and the argument from ontological non-fundamentalism only implies a non-reductivism that has to be distinguished from a stronger anti-reductivism.

PCS is more ambitious in proposing a positive account of the uniqueness of phenomenal concepts that is supposed to show *why* reductive explanations are not a tenable goal in the case of consciousness. PCS usually builds in some way on the observation that the deployment of phenomenal concepts requires acquaintance with phenomenal states, while the deployment of non-phenomenal concepts does not require acquaintance with the states to which they refer.[5] In order to successfully deploy a phenomenal concept, it is necessary to experience the phenomenal state or to have experienced it before. In order to successfully deploy a non-phenomenal concept, it is not necessary to experience the non-phenomenal state or have experienced it before.

The assumption that phenomenal concepts presuppose acquaintance can be further motivated in the context of classical thought experiments in philosophy of mind. Returning to the example of Mary, the design of the thought experiment implies that Mary knows everything about color vision from a third person perspective, but that she is not *acquainted* with the perception of color. If the deployment of phenomenal concepts presupposes acquaintance with the states they refer to, Mary cannot successfully deploy phenomenal color concepts before she leaves her prison. The assumption that the deployment of phenomenal concepts requires acquaintance with phenomenal states is equally helpful for other arguments in the debate about phenomenal consciousness. Consider Nagel's bats. If the deployment of phenomenal concepts presupposes acquaintance with the phenomenal states they refer to, then there is an obvious reason why the phenomenal perspective of bats remains opaque. We are not bats and therefore we are not acquainted with the phenomenal states of bats. Finally, consider the conceivability of zombies. If phenomenal concepts require acquaintance with their referents, then even a "complete" physical description of a conscious human being will not imply a phenomenal perspective, because even a complete physical description does not imply acquaintance with the described entities. Therefore, we can consistently imagine a physical duplicate of a conscious human being without any phenomenal states.

In moving beyond the basic idea of acquaintance-dependency, proponents of PCS propose a variety of specific accounts of phenomenal concepts. Loar's (1990) groundbreaking discussion presents phenomenal concepts in terms of a recognitional-demonstrative account, according to which phenomenal concepts differ from physical concepts in having the form "x is of *that* kind" (Loar 1997, 600). Imagine a headache and the thought "Oh no, not *that* headache again." According to Loar, the phenomenal concept "*that* headache" involves a demonstrative instead of a

---

[5] For the notion of acquaintance compare Russell (1912/2001, Chap. 5) and Balog's (2012) more recent discussion.

descriptive mode of presentation. The ability to refer to a specific headache is independent from the ability to describe it in a way that distinguishes it from other headaches. Instead, the phenomenal concept *"that* headache" seems to require the ability to demonstratively focus on the experience and to recognize different instantiations of the same type.

The recognitional-demonstrative character of phenomenal concepts contrasts with the theoretical character of physical concepts and Loar suggests that this difference already provides the first step in understanding the epistemic gap between the phenomenal and physical perspective: "What then accounts for the conceptual independence of phenomenal and physical-functional concepts? The simple answer is that recognitional concepts and theoretical concepts are in general conceptually independent" (Loar 1997, 602).

However, the recognitional-demonstrative character of phenomenal concepts cannot be the whole story as not every recognitional concept is a phenomenal concept. Consider, for example, someone using the non-phenomenal recognitional concept of *"that* dog". According to Loar, the crucial difference between *"that* headache" and *"that* dog" is that the latter is based on a "contingent mode of presentation" which means that the concept picks out its referent through contingent and therefore non-essential properties such as its visual appearance or its barking. Loar suggests that the situation is different in the case of phenomenal concepts: one can imagine the dog without these contingent properties, but one cannot imagine a headache without the phenomenal property of a headache feeling. Phenomenal concepts provide a direct grasp of their referents that is not mediated through a contingent property. "It is natural to regard our conceptions of phenomenal qualities as conceiving them as they are in themselves, i.e. to suppose we have a direct grasp of their essence" (Loar 1990, 608–09). To sum up, Loar's variant of PCS is based on two ideas. First, phenomenal concepts are recognitional concepts which partly explains their independence from physical concepts. Second, they are different from non-phenomenal recognitional concepts by not relying on a contingent mode of presentation but conceiving phenomenal qualities as they are in themselves.

Although Loar's recognitional-demonstrative proposal is widely recognized as the classical formulation of PCS, many current proponents of PCS prefer alternative accounts of phenomenal concepts. Michael Tye (2003), for example, argues that phenomenal concepts refer via the causal connection they have with their referents and takes his proposal to offer an explanation of the crucial epistemic features of the phenomenal perspective.[6] The quotational model of phenomenal concepts (e.g. Balog 2012) goes even further and argues that phenomenal concepts are not caused but partly constituted by the phenomenal states they refer to.

Both Tye's causal account and the quotational model can be understood as further developing parts of Loar's proposal by offering accounts of crucial features of the recognitional-demonstrative proposal such as the direct reference and the non-contingent mode of presentation of phenomenal concepts. However, not all current accounts of PCS are that close to Loar's proposal. David Papineau (2006),

---

[6] Tye (2010) now rejects PCS and argues that there are no phenomenal concepts.

for example, has argued that phenomenal concepts are not demonstrative but should be understood as retrieval of stored sensory templates. According to Papineau, sensory templates are set up on initial perceptual encounters with their referents. They can be reactivated on later occasions such as encounters with the same referents or in imagination. Papineau suggests that phenomenal concepts are also based on sensory templates: "I want now to suggest that we think of phenomenal concepts as simply a further deployment of the same sensory templates, but now being used to think about perceptual experiences themselves, rather than about the objects of those experiences. I see a bird, or visually imagine a bird, but now I think, not about that bird or a species, but about the experience, the conscious awareness of a bird" (Papineau 2006, 122).

8.4 Phenomenal Concepts and Physicalism

Despite obvious similarities between current accounts of PCS and conceptual pluralism, there is a crucial difference. While most proponents of PCS are physicalists, conceptual pluralists insist that we do not need to reduce all ontologies to a fundamental physical ontology. In this section, I argue that this additional assumption of a fundamental physical ontology renders PCS unsuccessful in the hands of physicalists, but successful in the hands of conceptual pluralists. An account of phenomenal concepts that is strong enough to justify non-reductivism will leave no room for the assumption of exactly one fundamental physical ontology while an account that is compatible with physicalism will be too weak to solve the problems of the explanatory gap.

In order to justify this claim, it will be helpful to have a closer look at common objections against PCS. Among the most influential critics of PCS are Terry Horgan and John Tienson, who argue that "new wave materialism" (i.e. a physicalist interpretation of PCS) is almost trivially self-defeating. Here is what they call the "deconstructive argument":

1. When a phenomenal property is conceived under a phenomenal concept, this property is conceived otherwise than as a physical-functional property.
2. When a phenomenal property is conceived under a phenomenal concept, this property is conceived directly, as it is in itself.
3. If (i) a property P is conceived, under a concept C, otherwise than as a physical-functional property, and (ii) P is conceived, under C, as it is in itself, then P is not a physical-functional property.

Hence,

4. Phenomenal properties are not physical-functional properties (Horgan and Tienson 2001, 311)

Chalmers (2006) offers a related argument that is based on the distinction between a "thin" and a "thick" phenomenal concept strategy: either phenomenal concepts are physically explicable or they are not physically explicable. If phenomenal concepts are physically explicable, then phenomenal concepts will be too "thin" to explain our phenomenal perspective and to demystify the explanatory gap. If phenomenal concepts are "thick" in the sense that they are not physically explicable, they will not be compatible with physicalism.

Chalmers presents this dilemma along the following lines. Let P be a complete microphysical account of the world and Q an arbitrary phenomenological truth. According to explanatory gap arguments, Q is not derivable from P and in this sense P&~Q is conceivable. PCS accepts that P&~Q is conceivable, but tries to explain this epistemic situation by pointing out the uniqueness of phenomenal concepts. Let C be a theory about whatever makes phenomenal concepts unique. According to Chalmers, we now have two options. On the one hand, we can endorse a thin account of C, according to which C is derivable from P and P&~C is not conceivable. On the other hand, we can insist on a thick account of phenomenal concepts, according to which C is not derivable from P and P&~C is conceivable. Chalmers argues that this creates a highly uncomfortable situation for physicalists who embrace the phenomenal concept strategy: if P&~C is not conceivable, then C will not explain the uniqueness of our phenomenal perspective and leave a residual explanatory gap. If P&~C is conceivable we create a new explanatory gap with respect to phenomenal concepts; we demystify phenomenal states at the price of mystifying phenomenal concepts. How can new wave materialists react to these objections? Following Chalmers' terminology, it is helpful to distinguish between two strategies.

Thin Accounts In the case of Horgan and Tienson's deconstructive argument, proponents of PCS can reject the second premise of the argument and insist that only physical concepts are fundamental in the sense that they conceive reality as it is in itself. Although it is possible to reject the second premise of the deconstructive argument, Loar is clearly committed to its truth as he explicitly claims that phenomenal concepts conceive phenomenal states "as they are in themselves" (Loar 1997, 608–09). However, one may argue that this is a problem of Loar's proposal that can be avoided by embracing a different account of PCS. Recall that there is a large variety of accounts of phenomenal concepts including Tye's recognitional-causal proposal, the quotational model, and Papineau's theory of sensory templates. It is far from clear that all of these accounts of PCS are committed to the claim that phenomenal concepts allow us to conceive phenomenal states as they are in themselves. Instead, one may claim that only physical concepts are fundamental but that phenomenal concepts are still different from other concepts and therefore lead to a unique epistemic situation.

However, there is an important reason to be suspicious about this strategy. The rejection of the second premise of the deconstructive argument threatens to weaken PCS in a way that it becomes ineffective as a non-reductive approach to the problem of the explanatory gap. Loar's new wave materialism provides an attractive theory of phenomenal consciousness *precisely* because of its suggestion that phenomenal concepts are fundamental, as well. If phenomenal concepts allow us to conceive phenomenal states as they are in themselves, then we have a good reason to reject the expectation of a reductive explanation of the phenomenal perspective.

The phenomenal perspective is fundamental and there is no reason to expect it to be physically explicable or to worry about an explanatory gap.

If we take only physical concepts to be truly fundamental, however, we undermine this deflationary strategy and it becomes unclear whether PCS still has anything new or interesting to say about the irreducibility of the phenomenal perspective. A variant of PCS that rejects the fundamentality of phenomenal concepts faces the same challenges as more traditional variants of non-reductive physicalism. If a proponent of PCS claims that phenomenal states are conceived as they are in themselves *only* if they are conceived under physical concepts, then we are again left with the question why there isn't a physical explanation of the phenomenal perspective.

An analogous problem is apparent in Chalmers' challenge of thin accounts of phenomenal concepts that suggest that the unique features of phenomenal concepts are physically explicable. Chalmers points out that this suggestion weakens PCS in an unacceptable way. If C is physically explicable, it seems obvious that C cannot explain our phenomenal perspective. In terms of Chalmers' conceivability considerations, if C is derivable from P, then C&~Q must be conceivable. Otherwise, we would have to deny the conceivability of P&~Q, which would be tantamount to the rejection of the explanatory gap. But if C&~Q remains conceivable then C does not explain our phenomenal perspective and does not do anything to dissolve the puzzlement over the explanatory gap. Thin phenomenal concepts won't help since the original problems, such as the conceivability of zombies, will reappear even under the assumption of C. We can imagine a duplicate of a conscious human being that satisfies whatever C requires and is still without consciousness.

Thick Accounts If "thin accounts" of phenomenal concepts are not sufficient to dissolve the puzzlement about the explanatory gap, proponents of PCS can attempt to justify a more ambitious account of phenomenal concepts. In the case of Horgan and Tienson's destructive argument, proponents of PCS can accept the second premise but challenge the third premise according to which fundamental phenomenal concepts would refer to fundamental phenomenal (and therefore non-physical) entities.

Horgan and Tienson do not consider this a viable strategy and even argue that the third premise is "virtually tautologous" (Horgan and Tienson 2001, 311). And indeed, if the phrase "conceiving an entity as it is in itself" is understood in terms of a strong metaphysical or ontological realism, the third premise seems to be a tautology. Either the fundamental ontology includes only physical entities or it also includes phenomenal entities. If the fundamental ontology only includes physical entities, then only physical concepts allow us to conceive an entity as it is in itself and thick accounts of phenomenal concepts are trivially wrong. If phenomenal concepts are fundamental in the sense that they also allow us to conceive an entity as it is in itself, then our fundamental ontology also includes phenomenal entities and physicalism is wrong. In other words: we have to choose between a thick account of phenomenal concepts and physicalism.

Although Horgan and Tienson do not discuss general metaphysical and ontological issues in their presentation of the deconstructive argument, Horgan has endorsed this kind of "metaphysical realism" in earlier publications (e.g. 1991). Horgan takes his "metaphysical realism" to be committed to the idea that "the only correct way of carving would be the one that corresponds to how THE WORLD is in itself— that is, the carving that picks out the genuine, mind-independently real, OBJECTS, and that employs predicates expressing the genuine, mind-independently real" (Horgan and Timmons 2002, 88).

Horgan's metaphysical realism has indeed serious implications for PCS: either phenomenal concepts carve the WORLD as it is in itself or they do not carve the WORLD as it is in itself. If they carve the WORLD as it is in itself, then physical concepts cannot conceive phenomenal states as they are in themselves and physicalism is wrong. If they do not carve the WORLD as it is in itself, then they cannot be fundamental and PCS must be wrong. In other words: given the commitment to metaphysical realism, it is indeed true that the third premise of the deconstructive argument is "virtually tautologous."

Chalmers also considers thick accounts of phenomenal concepts to be incompatible with physicalism. A thick account of PCS might, for example, point out the acquaintance-dependency of phenomenal concepts, without claiming that acquaintance is physically explicable in terms of some neural mechanism. Or, one could endorse the constitutional view according to which phenomenal concepts are at least partly constituted by phenomenal states (cf. Balog 2012), but resist the assumption that this feature must be explained on a more fundamental neural or physical level.

Chalmers endorses this strategy but insists that it is incompatible with physicalism.[7] According to Chalmers, the thick variant of PCS provides an answer to traditional formulations of the explanatory gap argument. For example, one can claim that phenomenal states are not physically explicable because the deployment of phenomenal concepts presupposes the acquaintance with the states to which they refer. However, if phenomenal concepts are not physically explicable, physicalists face what Chalmers calls a "second-order explanatory gap." Where physicalists used to be troubled by the existence of apparently fundamental phenomenal states, they should now be troubled by the existence of apparently fundamental phenomenal concepts. "[E]ven if there is a sort of explanation of the explanatory gap in terms of features of phenomenal concepts, the explanatory gap recurs just as strongly in the explanation of phenomenal concepts themselves." (2010, 321). For example, if the explanatory gap is due to the acquaintance-dependency of phenomenal concepts, we seem to end up with an inexplicable and therefore mysterious acquaintance relation. As proponents of a thick phenomenal concept strategy argue that the unique features of phenomenal concepts are not physically explicable, we seem to demystify the traditional explanatory gap at the price of mystifying phenomenal concepts.

---

[7] See Chalmers (2010, Chap. 8) for the endorsement of the constitutional account and (2010, Chap. 10) for the rejection of physicalist interpretations of the phenomenal concept strategy.

Beyond the dilemma The distinction between thin and thick accounts of phenomenal concepts leads to a dilemma for proponents of PCS. The first horn is a thin account of phenomenal concepts that is too weak to constitute an interesting answer to explanatory gap problems. The second horn is a thick account that turns out to be incompatible with physicalism.

Conceptual pluralists will embrace the second horn of the dilemma and a thick account of phenomenal concepts. In the case of the deconstructive argument, a conceptual pluralist will not claim that only physical concepts are truly fundamental but rather challenge the notion of "conceiving entities as they are in themselves" as presupposed in Horgan and Tienson's argument. If we do not assume that there is only one correct was of conceiving reality, we also do not have to decide whether phenomenal concepts or physical concepts allow us to conceive phenomenal states as they are in themselves. Instead, *both* phenomenal and physical concepts can be fundamental in the sense that they allow us to conceive the same reality in terms of different but equally fundamental ontologies.

Analogous considerations apply to Chalmers' argument against PCS. Chalmers' argues that a thick account of phenomenal concepts dissolves the traditional explanatory gap at the price of creating a new explanatory gap. By accepting the conceivability of P&~C, physicalists face a second-order explanatory gap that leaves the relation between phenomenal and physical concepts puzzling. In the end, physicalists do not dissolve but relocate the explanatory gap. Even if all of this is correct, however, conceptual pluralists will not be troubled. The very point of conceptual pluralism is to consider the scope of reductions and reductive explanations an open empirical question by allowing a plurality of equally fundamental ontologies. Even if C explains the conceivability of P&~Q, there is no reason to assume that there must be a physical account of C that would leave P&~C inconceivable.

The dilemma of a thick and a thin account of phenomenal concepts therefore illustrates why PCS is successful in the hands of conceptual pluralists even if it is unsuccessful in the hands of physicalists. I have introduced PCS as the claim that explanatory gap is not due to metaphysical differences between phenomenal and physical *states* but differences between phenomenal and physical *concepts*. The rejection of a substantive metaphysical gap renders PCS an attractive strategy for non-reductive physicalists. Unfortunately there are good reasons to doubt that this strategy can be successful: as soon as the appeal to different concepts is combined with the ideal of one fundamental physical ontology, the entire strategy becomes unstable. On the one hand, we can employ a strong notion of fundamental phenomenal concepts that seems to be incompatible with a physicalist framework because it adds fundamental phenomenal entities to our ontology. On the other hand, we can also employ a weaker account of phenomenal concepts that accepts the priority of physical concepts and insists that a fundamental ontology only includes physical entities. However, it then becomes unclear how weak phenomenal concepts can dissolve any physicalist worries about the explanatory gap.

Conceptual pluralism avoids the problems of physicalist proponents of PCS by rejecting the very idea of one fundamental ontology. A conceptual pluralist can accept that both phenomenal and physical concepts are fundamental because there is not just one fundamental way of "conceiving entities as they are in themselves". At the same time, conceptual pluralism preserves the non-dualist core idea of PCS that the explanatory gap is due to difference between phenomenal and physical concepts and does not imply a substantive metaphysical gap between the phenomenal and physical. Phenomenal and physical concepts may lead to mutually irreducible phenomenal and physical ontologies. However, this ontological pluralism has to be distinguished from a strong metaphysical pluralism that understands a plurality of ontologies in terms of a plurality of metaphysically distinct realms of reality.

To sum up: Even if Horgan and Tienson's destructive argument and Chalmers' dilemma are successful in undermining "New Wave Materialism", PCS may still provide an attractive framework for a "New Wave Pluralism" that combines an account fundamental phenomenal concepts with pluralism instead of physicalism (Ludwig 2013).

References

Balog, Katalin. 2012. Acquaintance and the Mind-body Problem. In *New Perspectives on Type Identity: The Mental and the Physical,* eds. Simone Gozzano and Christopher S Hill, 16–42. Cambridge: Cambridge University Press.

Block, Ned. 2002. The Harder Problem of Consciousness. *The Journal of Philosophy* 99 (8): 391–425.

Block, Ned, and Robert Stalnaker. 1999. Conceptual Analysis, Dualism, and the Explanatory Gap. *Philosophical Review* 108 (1): 1–46.

Bogardus, Tomas. 2013. Undefeated dualism. *Philosophical studies* 165 (2): 445–466.

Chalmers, David. 1995. Facing up to the problem of consciousness. *Journal of consciousness studies* 2 (3): 200–219.

Chalmers, David. 2006. Phenomenal Concepts and the Explanatory Gap. In *Phenomenal Concepts and Phenomenal Knowledge: New Essays on Consciousness and Physicalism*, 167–94. Oxford: Oxford University Press.

Chalmers, David. 2010. *The Character of Consciousness*. Oxford: Oxford University Press.

Dennett, Daniel C. 1993. *Consciousness Explained*. Boston: Little, Brown and Co.

Eronen, Markus I. 2011. *Reduction in Philosophy of Mind: A Pluralistic Account*. Berlin: Walter de Gruyter.

Faye, Jan. 2014. *The Nature of Scientific Thinking: On Interpretation, Explanation and Understanding*. New York: Palgrave Macmillan.

Horgan, Terence. 1991. Metaphysical Realism and Psychologistic Semantics. *Erkenntnis* 34 (3): 297–322.

Horgan, Terry, and John Tienson. 2001. Deconstructing New Wave Materialism. In *Physicalism and Its Discontents,* eds. Carl Gillett and Barry Loewer, 307–18. Cambridge: Cambridge University Press.

Horgan, Terry, and Mark Timmons. 2002. Conceptual Relativity and Metaphysical Realism. *Noûs* 36 (1): 74–96.

Hüttemann, Andreas. 2003. *What's Wrong with Microphysicalism?* New York: Routledge.

Jackson, Frank. 1982. Epiphenomenal qualia. *The Philosophical Quarterly* 32 (127): 127–136.

Loar, Brian. 1990. Phenomenal States. *The Nature of Consciousness: Philosophical Debates*, eds. Ned Block et al., 597–616. Cambridge, Mass.: MIT Press.

Loar, Brian. 1997. Phenomenal States (Second Version). In *The Nature of Consciousness: Philosophical Debates* eds. Ned Joel Block, Owen J. Flanagan, Güven Güzeldere, 597–615. Cambridge, Mass.: MIT Press.

Ludwig, David. 2013. New Wave Pluralism. *Dialectica* 67 (4): 545–60.

Levin, Janet. 2008. Taking Type-B Materialism Seriously. *Mind and Language* 23 (4): 402–25.

Macdonald, Cynthia, and Graham Macdonald. 2010. Emergence and Downward Causation. In *Emergence in Mind,* eds. Graham Macdonald; Cynthia Macdonald, 139–168, Oxford: Oxford University Press.

McLaughlin, Brian P. 2001. In Defense of New Wave Materialism: A Response to Horgan and Tienson. In *Physicalism and Its Discontents,* eds. Carl Gillett and Barry Loewer, 307–18. Cambridge: Cambridge University Press.

Nagel, Thomas. 1974. What Is It Like to Be a Bat? *The Philosophical Review* 38 (4): 435–50.

O'Connor, Timothy and Wong, Hong Yu. 2005. The Metaphysics of Emergence. *Noûs* 39: 658–678.

Papineau, David. 2006. Phenomenal and Perceptual Concepts. In *Phenomenal Concepts and Phenomenal Knowledge: New Essays on Consciousness and Physicalism,* 111–144. Oxford: Oxford University Press.

Polger, Thomas W. 2008. H2O, 'Water', and Transparent Reduction. *Erkenntnis* 69 (1): 109–130.

Popper, Karl Raimund. 1978. *Three Worlds. The Tanner Lecture on Human Values.* Minneapolis: University of Michigan.

Russell, Bertrand. 1912. *The Problems of Philosophy.* Oxford: Oxford University Press.

Sartenaer, Olivier. 2013. Neither Metaphysical Dichotomy nor Pure Identity: Clarifying the Emergentist Creed. *Studies in History and Philosophy of Science Part C: Studies in History and Philosophy of Biological and Biomedical Sciences* 44 (3): 365–373.

Sartenaer, Olivier. forthcoming. Disentangling the Vitalism–Emergentism Knot Wolfe, C.T. & Manda C.A. (eds.): *Forms of vitalism: Contemporary Metaphysics of Life and Scientific Intimations.* London: Pickering and Chatto.

Shear, Jonathan. 1999. *Explaining Consciousness: The Hard Problem.* Cambridge, Mass.: MIT Press.

Tye, Michael. 2003. A Theory of Phenomenal Concepts. *Royal Institute of Philosophy Supplement* 53: 91–105.

Tye, Michael. 2010. *Consciousness revisited: Materialism without phenomenal concepts.* Cambridge, Mass.: MIT Press.

Webster, William Robert. 2006. Human zombies are metaphysically impossible. *Synthese* 151 (2): 297–310.

Wong, Hong Yu. 2010 The Secret Lives of Emergents. *Emergence in Science and Philosophy,* eds. Antonella Corradini and Timothy O'Connor; 7–46. New York: Routledge.

Chapter 9
Beyond Dualism and Physicalism

I have presented conceptual pluralism as an alternative to both physicalism and dualism. For example, I have argued that PCS is successful in the hands of conceptual pluralists but not successful in the hands of physicalists who are committed to a strong metaphysical interpretation of the priority of the physical. Furthermore, I have argued that such a pluralist interpretation of PCS does not imply dualism as it does not require a metaphysical gap between the mental and the physical.

One may worry that this is not a stable position and that every account of conceptual pluralism will eventually collapse into either dualism or physicalism. Although I think that there are good reasons to distinguish conceptual pluralism from both dualism and physicalism, it is also important to acknowledge the vagueness of both terms "dualism" and "physicalism". Certainly, we could broaden the terms "dualism" and/or "physicalism" enough to include conceptual pluralism. I have no principled objections against this, as I do not believe that the diverse uses of both labels in metaphysics, philosophy of mind, and philosophy of science all express the same philosophcial claims. Still, I think that it is helpful to distinguish "conceptual pluralism" from both terms in philosophy of mind, as conceptual pluralism turns out to be incompatible with core claims of both dualists and physicalists in philosophy of mind.

9.1 But Isn't This Dualism?

In discussing the relation between conceptual pluralism and dualism, it is instructive to first have a general look at pluralism. Of course, there is a trivial difference between dualism and pluralism as dualists assume two types of entities (e.g. physical and mental) while pluralists assume more than two types of entities. Clearly, this is not sufficient to draw a helpful line between pluralism and dualism in philosophy of mind. To illustrate this point, consider Popper's pluralism of "three worlds"

© Springer International Publishing Switzerland 2015
D. Ludwig, *A Pluralist Theory of the Mind*, European Studies
in Philosophy of Science 2, DOI 10.1007/978-3-319-22738-2_9

(1978) that not only includes a physical world and a mental world but also a third world of objective knowledge. Still, it would be odd to claim that Popper is not a dualist as his pluralism includes a traditional interactionist dualism in philosophy of mind (e.g. Popper and Eccles 1977). Furthermore, it is easy to clarify this situation by specifying that Popper's general pluralist metaphysics entails a *mind-body* dualism. In other words: Popper's commitment to the existence of non-physical and non-mental entities is by no means in conflict with traditional dualist accounts of the relation between physical and mental entities.

While Popper's pluralism of "three worlds" clearly implies a mind-body dualism, I have argued that conceptual pluralism should be distinguished from metaphysically ambitious forms of pluralism. Although conceptual pluralism implies a plurality of ontologies, this ontological pluralism is built on conceptual relativity and the idea that different conceptual choices imply different ontologies. We can describe the same reality in terms of different conceptual frameworks and therefore do not need to assume that mutually irreducible ontologies come with a substantive form of metaphysical distinctness (cf. Sect. 10.1 for a discussion of "metaphysical distinctness").

In arguing for this idea of conceptual relativity in ontology, I have relied on examples from contemporary metaphysics and scientific practice. In metaphysics, Putnam's thought experiment of three elementary particles in an empty space provides a helpful example. According to conceptual relativity, there is not one fundamental ontology as we can describe reality in terms of different but equally fundamental conceptual frameworks. For example, we can describe Putnam's toy universe in terms of a nihilist or a universalist framework that imply different ontologies. Still, it would be very odd to claim that nihilist and universalist ontologies refer to metaphysically distinct entities in any metaphysically ambitious sense. The same point can be made with the plurality of ontologies that we find in scientific practice. For example, I have argued that different explanatory interests in biology lead to different biological ontologies as illustrated by different accounts of species. Again, it would be odd to interpret this plurality of biological ontologies in a metaphysically ambitious sense as it seems uncontroversial that different accounts of species conceptualize the same biological reality in terms of different explanatory interests.

If we extend this conceptual pluralism to philosophy of mind, we again end up with a metaphysically shallow interpretation of the differences between physical and phenomenal ontologies. While we can describe humans in terms of very different (e.g. physical, biological, psychological) ontologies, this ontological plurality is not sufficient for a strong account of metaphysical distinctness as it is assumed by dualists. Even if these ontologies are mutually irreducible, we can point out that a plurality of equally fundamental ontologies is common both in metaphysics and the empirical sciences and does by no means require a traditional dualist framework.

Of course, one could object that conceptual pluralism does not constitute a traditional form of dualism but still constitutes *some* form of dualism as it comes with the assumption of irreducible non-physical ontologies. At this point, it is important to resist the temptation to engage in a verbal dispute. Of course, we can define "dualism" in a way that it trivially includes conceptual pluralism but I think that such as definition would not be helpful as it would blur important lines in the debate. An overly liberal definition of dualism would create more confusion than clarity as it would trivialize the distinctness of mental and physical states in a way that contradicts the intention of philosophers who usually consider themselves as dualists. Clearly, most dualists in philosophy of mind do not want to claim that mental and physical entities are distinct in a similar way as nihilist and universalist objects or as ecological and morphological species. Instead, they aim at a much more substantive sense of metaphysical distinctness, which suggests that we should clearly distinguish between dualism and conceptual pluralism. There may be other contexts and other reasons to prefer a very liberal notion of "dualism" that includes conceptual pluralism. I do not mean to police how people use the label "dualism". If there are contexts in which it is helpful to describe conceptual pluralism as "dualism," that's fine. I just don't think that it's helpful to call conceptual pluralism "dualism" in debates about the metaphysics of mind as most philosophers will expect more than the metaphysically shallow pluralism that I have to offer.

9.2 But Isn't This Physicalism?

Given the assumption that conceptual pluralism should be distinguished from dualism, one may wonder whether conceptual pluralism turns out to be some form of physicalism. The discussion of the phenomenal concept strategy (PCS) has already revealed a number of important similarities between conceptual pluralism and non-reductive physicalism and I anticipate the objection that my distinction between conceptual pluralism and physicalism has been premature. More specifically, one may insist that the rejection of any substantive metaphysical gap between the mental and the physical is already sufficient for some minimal notion of physicalism.

However, there are good reasons to argue that the rejection of metaphysical distinctness is not sufficient for physicalism as physicalism also requires a substantive notion of the ontological priority of the physical. At best, the rejection of metaphysical distinctness establishes some minimal notion of monism. However, monism comes in at least three flavors: physicalism, neutral monism, and idealism. While all three types of monism can agree on the rejection of metaphysical distinctness, they disagree on issues of priority. Physicalists insist on the priority of the physical, idealists insist on the priority of the mental, and neutral monists reject both priority claims.

In order to present conceptual pluralism as variant of physicalism (instead of, for example, neutral monism), we would therefore need some robust notion of ontological priority. Unfortunately, my discussion of the argument from ontological non-fundamentalism (Sects. 7.1, 7.2, and 7.3) casts doubts on the availability of such a notion. Although I have discussed a variety of modest notions of priority such as composition, scope of application, or nomological supervenience, none of them turned out to be robust enough to justify some form of physicalism.

Arguably, the most promising strategy to defend a physicalist pluralism is based on the notion of metaphysical supervenience. In the discussion of the argument from ontological non-fundamentalism, I have suggested that metaphysical supervenience will not provide a notion of ontological priority that is strong enough to justify reductivism. However, it may still be strong enough to justify non-reductive physicalism and therefore lead to a happy coexistence of conceptual pluralism and non-reductive physicalism: while conceptual pluralism explains why we should not expect a reductive explanation of the mind, metaphysical supervenience explains why we should still endorse a moderate notion of the ontological priority of the physical.

The most common pluralist reaction to this suggestion is an outright rejection of supervenience claims (e.g. Dupré 1993; Putnam 1999; Horst 2007). And indeed, it is not difficult to raise doubts about supervenience claims within a pluralist framework. First, it is far from clear that the claim that *everything* supervenes on the physical is actually supported by our empirical knowledge about the world and that an induction from well-established case studies to a general supervenience thesis is possible. This problem seems especially pressing in the light of various forms of externalism that suggest that psychophysical supervenience often has to take the form of some speculative "global supervenience" (cf. Leuenberger 2009; Steinberg 2014) instead of an empirically verifiable form of local psychoneural supervenience. In the end, generalized supervenience claims may therefore turn out to be a metaphysical postulate that is not backed up by our empirical knowledge about the world and that is only motivated by the intuitions of physicalist philosophers.

My discussion of conceptual pluralism, however, suggests a different strategy as even a stable notion of metaphysical supervenience would not be sufficient to turn conceptual pluralism into a variant of non-reductive physicalism (cf. Daly 1997). Consider my brief discussion of "necessitation dualism" and "necessitation monism" in the context of the argument from ontological non-fundamentalism in Sect. 7.2. The "necessitation dualist" has been introduced by Stoljar (2010) as a hypothetical dualist who insists that physical and mental properties are metaphysically distinct but still accepts that mental properties metaphysically supervene on physical properties. If necessitation dualism is a consistent position, metaphysical supervenience is clearly not sufficient for physicalism.

I have added a further aspect to this argument by pointing out that dualism is not the only metaphysical alternative to physicalism. We can also imagine a "necessitation monist" who rejects physicalism by refusing to accept the priority of the physical in the same sense as she rejects idealism by refusing to accept the priority of the mental. Note that at this point the question is not so much whether necessitation monism is true or plausible but rather whether it is a coherent position. If we want to formulate physicalism in terms of metaphysical supervenience, we have to exclude all coherent nonphysicalist alternatives, no matter whether they turn out to be true in the end.

Furthermore, I have suggested that the tradition of "psychophysical parallelism" that stretches from Gustav Fechner's *Elements of Psychophysics* (1860) to Feigl's *The "Mental" and the "Physical"* (1967) provides a good example of necessitation monism. Recall that Moritz Schlick, one of the most prominent proponents of this position, presented "psychophysical parallelism [as] a harmless parallelism of two differently generated concepts" (Schlick 1927). Schlick insists that his psychophysical parallelism implies that mental and physical states are *not* metaphysically distinct as it is only "an epistemological parallelism between a psychological conceptual system on the one hand and a physical conceptual system on the other. The 'physical world' *is* just the world that is designated by means of the system of quantitative concepts of the natural sciences." (1918/1974, 301). Whereas dualists assume an metaphysical gap, Schlick argues for a conceptual gap. "There is only *one* reality," (1918/1974, 244) but this reality can be described in terms of different conceptual systems.

While all of this sounds quite familiar to non-reductive physicalists and especially to proponents of the phenomenal concept strategy, Schlick is also very vocal in his rejection of materialism as he argues that physical concepts should not be considered metaphysically prior and that materialists are as wrong as idealists who consider only the mental to be fundamental. "Earlier we were obligated most emphatically to reject the mistaken idea that a different kind or a different degree of reality must be ascribed to these two groups of reality [the mental and the extramental], that one group is to be characterized as merely an "appearance" of the other. On the contrary, they are all to be regarded as, so to speak, of equal value." (Schlick 1918/1974, p. 244).

If necessitation monism is a consistent position, metaphysical supervenience will be of no help in justifying the ontological priority of the physical and in formulating a variant of physicalism that is compatible with conceptual pluralism. To sum up, we end up with the following simple argument against physicalist interpretation of conceptual pluralism: physicalism requires a substantive notion of the ontological priority of the physical. Conceptual pluralism does not leave room for a substantive notion of the ontological priority of the physical. Therefore, conceptual pluralism is not a variant of physicalism.

One may object this argument sets the bar for "physicalism" too high. For example, Eronen (2011) has argued for a "pluralistic physicalism" despite his acknowledgment of ontological plurality. In addition to supervenience considerations, Eronen points out that one "should not understand ontological pluralism [...] as some kind of 'spooky' pluralism that asserts that there are fundamentally different substances in the world." (150). Furthermore, he refers to Ladyman and Ross' (2007) *Primacy of Physics Constraint* according to which "physics sets *constraints* for the theories of special sciences" (151).

At this point, it is again important to point out that debates about "physicalism" and "dualism" can easily become verbal disputes. Of course, no one can stop us from defining physicalism as a largely uncontroversial thesis about the nonexistence of "spooky substances" even if this would turn a good number of contemporary property dualisms into physicalism. Furthermore, we can also define "physicalism" as purely methodological thesis such as the *Primacy of Physics Constraint* and may even point out that physicalism was originally conceived by logical positivists as a methodological thesis. Finally, we can also define "physicalism" through supervenience theses, even if they include monists like Schlick who explicitly reject the priority of the physical.

Again, the point is not to police how people use labels and there may be contexts in which it is helpful to describe conceptual pluralism as "physicalism". That's fine with me. Still, it seems to me that the label "physicalism" would be mostly misleading in current debates about the metaphysics status of the mind. Most physicalist philosophers of mind do not only want to get rid of "spooky substances" or dubious methodologies but they want to make a metaphysical claim that clearly distinguishes them from neutral monists, idealists, positivists, and so on. An overly liberal definition of physicalism that includes conceptual pluralism therefore seems to create more confusion than clarity in current debates of philosophy of mind.[1]

9.3 The Identity Objection

Contrary to physicalism, conceptual pluralism rejects the ontological priority of the physical. *Contrary to dualism*, conceptual pluralism rejects the idea that irreducibility implies any substantive form of metaphysical distinctness. I anticipate the objection that conceptual pluralism does not constitute a stable alternative and will collapse into either physicalism or dualism. I have argued that phenomenal and physical states are not metaphysically distinct in any substantive sense and therefore do not support dualism. However, one may object that this formulation avoids the crucial question whether phenomenal and physical concepts refer to the *same* entities. Furthermore, one can suggest that any answer to this question will expose conceptual pluralism as either a version of physicalism or dualism. If conceptual pluralists accept that phenomenal and physical concepts refer to the same entities, conceptual pluralists will also have to accept at least some minimal version of token physicalism. If they argue that phenomenal and physical concepts do not refer to the

---

[1] It is important to note the qualification that this is not what philosophers *of mind* usually mean with "physicalism". In philosophy of mind (and large parts of analytic metaphysics), physicalism is taken to imply property physicalism and is also taken to be incompatible with neutral monisms that reject the ontological priority of the physical. The situation is different in other philosophical subcommunities such as philosophy of biology where Rosenberg's claim "We're all physicalists now" (2008, 4) captures a widespread sentiment despite the influence of anti-reductionist and pluralist positions (e.g. Kaiser 2011, Brigandt and Love 2012). Arguably, this situation simply reflects different interests uses of "physicalism". First, philosophers of biology often equate physicalism with "token physicalism" and bracket the question of the ontological status of types that fuels debates in philosophy of mind. Philosophers of mind and metaphysicians are usually highly skeptical of this suggestion (cf. Kim 2012) and Stoljar summarizes the most obvious objection that "the token physicalism maneuver [...] is consistent with a standard form of dualism, viz. Property dualism."(2010, 121). Furthermore, philosophers of biology are usually not interested in abstract debates about "ontological priority" and may be quite happy to admit that their account of "token physicalism" does not exclude monisms in the tradition of Fechner and Schlick. Again, it seems to me that this is mostly a verbal issue that will be decided by pragmatic and rhetorical concerns.

same entity, conceptual pluralism will imply a dualist assumption of non-physical entities. Given these consideration, one can formulate the following "identity objection" against conceptual pluralism:

1. Conceptual pluralism has to either accept or reject the identity of mental and physical entities.
2. If conceptual pluralism accepts the identity of mental and physical entities, it collapses into physicalism.
3. If conceptual pluralism rejects the identity of mental and physical entities, it collapses into dualism.
∴ Conceptual pluralism either collapses into physicalism or dualism.

How can a conceptual pluralist react to this argument? Putnam defends his variant of pluralism by challenging the first premise: "the notion of identity has not been given any sense in this context. We cannot, for example (as I once thought we could), employ the model of theoretical identification derived from such famous successful reductions such as the reduction of thermodynamics to statistical mechanics, because that model assumes that both the reduced theory and the reducing theory have a well-defined body of laws" (1999, 85).

My strategy will be different and I will accept the first premise at least for the sake of the argument. However, I think that a conceptual pluralist has good reasons to reject both the second and third premise of the identity objection. The second premise is flawed because physicalism requires more than the identity of mental and physical entities. As I have argued in the previous sections, "physicalism" in its usual meaning in philosophy of mind and metaphysics requires the ontological priority of the physical. (Some) idealists claim that mental and physical entities are identical and that the mental entities are ontologically prior. (Some) neutral monists claim that mental and physical entities are identical and that the neither of them are ontologically prior. Therefore, identity claims are not enough for physicalism.

The rejection of the second premise can be illustrated with the historical examples from the introduction. The assumption of mind-body identity is not sufficient for physicalism since there are many historical positions that accept the identity claim without accepting the priority of the physical. Psychophysical parallelism in the tradition from Fechner to Feigl is one clear example of how identity assumptions have been combined with an explicitly non-materialist framework. Of course, we can also point to an idealist or "psychomonist" position that accepts mind-body identity, but combines it with the ontological priority of the mental.

Furthermore, we do not need historical examples to establish that the assumption of mind-body identity does not imply physicalism. Physicalism presupposes an asymmetric relation in which the physical turns out to be prior. As identity is a symmetric relation it can obviously not be sufficient for physicalism. This trivial fact is often ignored by philosophers who talk as if mind-body identity and physicalism were different labels for the same idea.

There is a simple explanation for the confusion about identity and physicalism: according to many physicalist theories of the mind, the identity of mental and physical states is established through reductions or reductive explanations. In the same way as we establish the identity of water and H_2O by explaining water in terms of H_2O, we are supposed to establish the identity of mental and physical states by explaining mental states in terms of physical states. Given this reductive model, mind-body identity and physicalism indeed come as package.

However, identities do not have to be established through theory reductions or reductive explanations, and they do not have to come with the ontological priority of one of the relata. Consider the identity of Molière and Jean Baptiste Poquelin, Hesperus and Phosphorus, or *Taraxacum officinale* and the common dandelion. These kinds of identities are not established through reductive explanations and they do not come with the ontological priority of one of the relata. Given that identities are not sufficient for physicalism, we should reject the second premise of the identity objection. A conceptual pluralist can accept that mental and physical entities are identical and still reject physicalism by rejecting the ontological priority of the physical.

9.4   Limits of Identity

Conceptual pluralists can accept identity claims without being committed to physicalism by rejecting the priority of one of the relata. However, I think that conceptual pluralists should actually go a step further and also reject the third premise of the identity objection. I other words, pluralists can reject the identity of mental and physical states without being committed to dualism. In the following, I will argue that there is a subtle but important difference between the rejection of metaphysical distinctness and the assumption of identity. Therefore, a conceptual pluralist can reject a metaphysical gap between the mental and the physical without having to accept the identity of mental and physical states.

The basic idea is easily illustrated by my various examples of conceptual relativity in philosophy and the empirical sciences. Again, consider the nihilist and universalist accounts of Putnam's universe with three individuals. While it seems obvious that the universalist's objects are not identical with the nihilits' objects, there is also no reason to assume that they describe distinct entities in any metaphysically substantive sense. Nihilists and universalists simply describe the same individuals in terms of different ontologies. The same considerations apply to my examples from the empirical sciences. Two psychiatric ontologies such as the *International Classification of Diseases* (ICD) and the *Diagnostic and Statistical Manual of Mental Disorders* (DSM) postulate psychiatric entities that can often not be identified but still leave no room for a substantive notion of metaphysical distinctness.

In order to clarify this distinction between identity and the negation of metaphysical distinctness, it is helpful to have a closer look at the concept of identity. Although we seem to have an intuitive grip of what "identity" means, it is surprisingly difficult to offer a satisfying definition. Of course, we can define identity as the relation each thing bears to itself and to nothing else. However, this definition is circular as "nothing else" presupposes that we already understand "non-identity" (cf. Noonan 2011).

The best-known and most traditional approach to identity is based on indiscernibility. There are two principles to consider here. On the one hand, the principle of the *identity of indiscernibles* states that x and y are identical if they are indiscernible. On the other hand, the principle of the *indiscernibility of identicals* states that x and y are indiscernible if they are identical. I will refer to the conjunction of both principles as *Leibniz Law*: x and y are identical if and only if x and y are indiscernible. Leibniz Law has the great advantage of not only capturing many intuitions about identity but also offering a usable criterion to establish identities: in order to figure out whether x and y are identical, we simply have to figure out whether x and y are indiscernible.

While Leibniz Law is clearly an attractive principle, it has surprising consequences for relations that we usually do not consider relations between metaphysically distinct entities. Given Leibniz Law, contemporary metaphysics is full of relations that seem to be neither cases of identity nor distinctness in a metaphysically substantive sense. Let us briefly consider three of them: composition, constitution, and determinables.

Composition   One of the most baffling consequences of Leibniz Law is that it casts doubt on the idea that composition entails identity. Consider a table and the atoms that compose the table. Is the table identical with the atoms that compose it? Given Leibniz Law, there are good reasons to doubt that composition implies identity. Most obviously, the table and the atoms that compose the table have different temporal properties. For example, it might be correct to say "This table is four years old" while it would be wrong to say "The atoms a_1–a_n are four years old." Given Leibniz Law, this seems to imply that the table and the atoms a_1–a_n cannot be identical. There is a further problem with the assumption that a table is identical with the atoms that compose it. The spatial boundaries of ordinary objects are not sharp enough to allow identification with a specific set of atoms. Consider the surface of a table. From a microphysical perspective, it simply is not possible to give a precise account of the boundaries of the table. Often, it won't be possible to say whether a specific atom is part of the table. As a consequence, we cannot identify the table with one specific set of atoms a_1–a_n.[2]

Constitution   Another potential example of the limits of identity is constitution. Consider a few cases of material constitution: a clay statue is constituted by a lump of clay, a copper coin is constituted by a piece of copper, and a ship is constituted by an arrangement of wooden planks. The first thing to notice is that there is a difference between composition and constitution. The lump of clay that constitutes the clay statue is not part of statue and it does not compose the statue. In the same way,

---

[2] Of course, this is a gross oversimplification of the complex debates about the relation between composition and identity (cf. McDaniel 2008). However, not much depends on this in the present context as my general point (it can be helpful to distinguish between identity and the rejection of metaphysical distinctness) does not depend on this example.

the piece of copper that constitutes the copper coin is not part of the copper coin and it does not compose the copper coin.

How, then, shall we understand constitution? First, one may argue that constitution is a form of identity. The lump of clay is not part of the clay statue but it *is* the statue. In the same way, the piece of copper is not part of the copper coin but it *is* the copper coin. However, the identity assumption faces obvious difficulties. Assume that a sculptor buys lump of clay and creates a statue of Goliath. However, she is unhappy with the result and remodels the lump into a statue of David. Given the assumption that the lump is identical with the statues and the transitivity of the identity relation, we face the following situation:

Lump = statue of Goliath
Lump = statue of David
∴ Statue of Goliath = statue of David

Obviously, this is an absurd result as the statue of Goliath is not identical with the statue of David. Given Leibniz Law, there are further reasons why we should not consider a statue to be identical with the lump of clay that constitutes the statue. For example, the statue and the lump differ in their temporal properties. Assume that the sculptor destroyed the statue of Goliath on New Year's Eve, 1900. However, she did not destroy the lump of clay but remodeled it into a statue of David. Therefore, the statue of Goliath has the property of having been destroyed on New Year's Eve, 1900, while the lump of clay does not have the property of having been destroyed on New Year's Eve, 1900. The statue of Goliath and the lump of clay also differ in their general persistence properties: the statue of Goliath cannot survive being squished while the lump of clay can survive being squished. Furthermore, it seems reasonable to argue that they also have different economic and aesthetic properties: for example, the statue of David might be worth 40.000$ while the lump of clay is only worth 40$.[3]

Determinables and Determinates   Let us consider a third and last example of limits of identity claims. In his 1921 *Logic,* W.E. Johnson proposed "to call such terms as color and shape determinables in relation to such terms as red and circular which will be called determinates." (Johnson 1921, 174). The general idea is that we can refer to features of objects in different ways. For example, we can say that an object is colored, that it is red, and that it is scarlet. In the same way, we can say that an object is angled, that it is square, or that it is quadratic. In this sense the relation between a determinable and determinate is a relation between a more general and a more specific way of referring to a feature of an object. "Colored" is a determinable of "red," "red" is a determinable of "scarlet," and so on. The argument for the non-identity of determinables and determinates is straightforward: a determinable has many determinates. For example the determinable red has the determinate rose,

---

[3] Again, this sketchy presentation does not do justice to the complex debates about the relation between constitution and identity (e.g. Rudder Baker 1997 and Wasserman 2014). Also, see Rudder Baker (2000) for an intriguing application of constitution to philosophy of mind.

scarlet, and Venetian red. If determinables and determinates were identical, we could falsely deduce that the different determinates are identical too:

Property of being red = property of being scarlet
Property of being red = property of being Venetian red
∴ Property of being scarlet = property of being Venetian red

The relation between determinables and determinates is less likely to be taken as an identity relation than composition or constitution. However, I think that all three types of relations share an important feature: even if they are not identity relations, it seems that they are also not relations between metaphysically distinct entities. In an important sense, there is not a table and *additionally* also the atoms that compose it. There is not a statue and *additionally* also lump of clay that constitutes it. There is not the property of being scarlet and *additionally* also the property of being red.

The cases of composition, constitution, and determinables/determinates provide further illustrations of the idea that it is often helpful to distinguish identity and the negation of metaphysical distinctness. There are many situations in which at least traditional accounts of identity fail and where we still want to say that we are not dealing with metaphysically distinct entities. To illustrate this "rejection of metaphysical distinctness", imagine a room with nothing but a table and a chair. Furthermore, imagine that the table is composed of the elementary particles e_1–e_n, while the chair is composed of the elementary particles e_o–e_p. Consider the following descriptions of the room:

1. There are a table and a chair
2. There are the elementary particles e_1–e_p
3. There are a table and the elementary particles e_o–e_p
4. There are a table and the elementary particles e_1–e_n
5. There are a chair and the elementary particles e_o–e_p

I think that there is an obvious sense in which (1)–(3) describe the entire room while (4) and (5) do not describe the entire room but leave parts of the room undescribed. We can describe the entire room in terms of ordinary language (1), microphysics (2), or a mixed vocabulary (3). Even if the entities of ordinary language and particle physics fail to meet traditional identity criteria, there is still an obvious sense in which we can describe the entire room in either of the conceptual frameworks as they do not refer to metaphysically distinct entities.

Analogous examples are possible in the case of material constitution. Consider a gallery room with a marble statue of Goliath, a clay statue of David, and nothing else. Again, we can describe the room in different ways.

1. There are a statue of Goliath and a statue of David
2. There are a lump of marble and a lump of clay
3. There are a statue of Goliath and a lump of clay
4. There are a statue of Goliath and a lump of marble
5. There are a statue of David and a lump of clay

Again, there is an important sense in which (1)–(3) describe the entire room while (4) and (5) only describe parts of it. This seems to be true even if a statue and the material that constitutes it fail to meet the identity criteria. I think that the best way to explain why (1)–(3) offer a description of the entire room even under the assumption of non-identity is to say that a statue and the material that constitutes it are not metaphysically distinct in a substantive sense.

There is another and perhaps even more powerful reason to argue that the rejection of identity does not imply metaphysical distinctness. Consider debates about mental causation and overdetermination (see next section). Often, it is assumed that identity assumptions are the only way to avoid overdetermination. By identifying two causes C1 and C2, we can avoid causal competition by arguing that C1 and C2 are in fact identical causes. A closer look at composition as well as determinables and determinates, however, suggests that failed identity claims do not imply overdetermination. Consider a work accident of a gallery owner. While trying to set up a new exhibition, the clay statue of David falls on her foot. She rushes to a doctor who asks her what happened to her foot. Here are two possible answers:

1. A lump of clay fell on my foot
2. A statue of David fell on my foot

What is the correct answer? Of course, both answers are correct. But if both answers are correct, how is this not a case of mysterious overdetermination? I think that the only plausible answer is that the lump of clay and the statue of David are not metaphysically distinct and therefore (1) and (2) do not describe metaphysically distinct causal processes. It was not a lump of clay and *additionally* also a statue of David that caused the injury. There was only one accident that can be described from different perspectives. A similar argument can be made in the case of determinables and determinates, as Stephen Yablo has shown:

> Imagine a pigeon, Sophie, conditioned to peck at red to the exclusion of other colors; a red triangle is presented, and Sophie pecks. Most people would say that the redness was causally relevant to her pecking, even that it was a paradigm case of causal relevance. But wait! I forgot to mention that the triangle is a specific shade of red: scarlet. (Yablo 1992, 257).

At first glance, Yablo's example of the determinate and determinable relation seems to be a case of overdetermination as there are two causes for the pigeon's behavior: the triangle being red and the triangle being scarlet. However, contrary to genuine cases of overdetermination, the causes do not compete with each other. It would be a misunderstanding to ask whether the scarletness or the redness caused Sophie to peck, because there is an obvious sense in which a determinable is not metaphysically distinct from its determinate. The triangle is not scarlet and *additionally* also red. Rather, "scarlet" specifies the redness of the triangle.[4]

---

[4] The idea that determinates and determinables provide a helpful model for the relation between mental and the physical properties has been criticized by many philosophers (e.g. Harbecke 2008, Haug 2010). However, it should be clear that my discussion does not depend on the assumption that mental properties *are* determinates. Instead, I use the example of determinates and determinables as an illustration of the claim that the rejection of identity does not always imply distinctness in a metaphysically ambitious sense that would create worries of overdetermination and dualism.

To sum up, identity seems to be too narrow to be equated with the rejection of metaphysical distinctness. Arguably, there are many cases of non-identity that do not imply metaphysical distinctness in the sense of dualist claims of the distinctness of mental and physical entities. Therefore a rejection of psychophysical identity claims is not sufficient for dualism and the third premise of the identity objection should be rejected.

References

Brigandt, Ingo and Love, Alan. 2012. Reductionism in Biology, *The Stanford Encyclopedia of Philosophy*, http://plato.stanford.edu/archives/sum2012/entries/reduction-biology

Daly, Chris. 1997. Pluralist metaphysics. *Philosophical studies* 87 (2): 185–206.

Dupré, John. 1993. *The Disorder of Things: Metaphysical Foundations of the Disunity of Science.* Cambridge, Mass.: Harvard University Press.

Eronen, Markus I. 2011. *Reduction in Philosophy of Mind: A Pluralistic Account.* Berlin: Walter de Gruyter.

Fechner, Gustav. 1860. *Elemente der Psychophysik.* Leipzig: Breitkopf und Härtel.

Feigl, Herbert. 1967. *The Mental and the Physical: The Essay and a Postscript.* Minneapolis: University of Minnesota Press.

Harbecke, Jens. 2008. *Mental Causation: Investigating the Mind's Powers in a Natural World.* Berlin: Walter de Gruyter.

Haug, Matthew C. 2010. Realization, determination, and mechanisms. *Philosophical Studies* 150 (3): 313–330.

Horst, Steven W. 2007. *Beyond Reduction: Philosophy of Mind and Post-reductionist Philosophy of Science.* Oxford: Oxford University Press.

Kaiser, Marie I. 2011. The Limits of Reductionism in the Life Sciences. *History & Philosophy of the Life Sciences*, 33 (4): 453–476.

Kim, Jaegwon. 2012. The Very Idea of Token Physicalism. In *New Perspectives on Type Identity: The Mental and the Physical,* eds. Simone Gozzano and Christopher S Hill, 167–193. Cambridge: Cambridge University Press.

Johnson, William E. 1921/1964. *Logic, Part 1.* New York: Dover Publications.

Ladyman, James, and Don Ross. 2007. *Every Thing Must Go: Metaphysics Naturalized.* Oxford: Oxford University Press.

Leuenberger, Stephan. 2009 What is global supervenience? *Synthese* 170 (1): 115–129.

McDaniel, Kris. 2008. Against Composition as Identity. *Analysis* 68 (298): 128–133.

Noonan, Harold. 2011. Identity. In *The Stanford Encyclopedia of Philosophy*, edited by Edward N. Zalta, Winter 2011. http://plato.stanford.edu/archives/win2011/entries/identity/

Popper, Karl Raimund. 1978. *Three Worlds. The Tanner Lecture on Human Values.* Minneapolis: University of Michigan.

Popper, Karl Raimund and John C. Eccles. 1977. *The Self and Its Brain.* Berlin: Springer.

Putnam, Hilary. 1999. *The Threefold Cord: Mind, Body, and World.* New York: Columbia University Press.

Rosenberg, Alexander. 2008. *Darwinian Reductionism: Or, How to Stop Worrying and Love Molecular Biology.* Chicago: University of Chicago Press.

Rudder Baker, Lynne. 1997. Why constitution is not identity. *The Journal of Philosophy* 94 (12): 599–621.

Rudder Baker, Lynne. 2000. *Persons and Bodies: A Constitution View.* Cambridge: Cambridge University Press.

Schlick, Moritz. 1918/1974. *General Theory of Knowledge*. Vienna: Springer.
Schlick, Moritz. 1927. Letter to Ernst Cassirer. Inv. No. 94. Schlick-Papers.
Steinberg, Alex. 2014. Defining Global Supervenience. *Erkenntnis* 79 (2): 367–380.
Stoljar, Daniel. 2010. *Physicalism*. New York: Routledge.
Wasserman, Ryan. 2014. Material Constitution, *The Stanford Encyclopedia of Philosophy,* eds. Edward N. Zalta. http://plato.stanford.edu/archives/sum2014/entries/material-constitution
Yablo, Stephen. 1992. Mental Causation. *The Philosophical Review* 101 (2): 245–80.

Chapter 10
Mental Causation

In the previous sections, I argued that conceptual pluralism allows us to take a relaxed attitude towards explanatory gaps and traditional puzzles in philosophy of mind. While a discussion of the unique features of phenomenal concepts may support anti-reductivism, conceptual pluralism ultimately takes a non-reductivist stance that considers the scope of reductive explanations an open empirical question. One way or another, there is no philosophical problem as long as we do not presuppose a strong but unnecessary notion of the ontological priority of the physical.

Even if we assume that this strategy is successful, one might still wonder whether conceptual pluralism is a stable approach to the philosophy of mind. For example, Jaegwon Kim has argued that there are actually *two* entangled mind-body problems: the problem of consciousness and the problem of mental causation (Kim 2005, 7). As Kim considers the problem of mental causation to be the crucial objection against non-reductive theories of the mind, it is reasonable to ask whether a pluralist theory of the mind can explain mental causation. As Kim puts it:

> I believe that the question no longer is whether those of us who want to protect mental causation find mind-body reductionism palatable. What has become increasingly clear after three decades of debate is that if we want robust mental causation, we had better be prepared to take reductionism seriously, whether we like it or not (Kim 2005, 22).

Why should we assume that only a reductive theory of the mind will provide a robust notion of mental causation? The problem is often presented along the following lines. We find causal explanations in a variety of different contexts such as physics, chemistry, biology, psychology, sociology, and so on. However, the "real causal powers" (cf. Glennan 2010; Pereboom 2002) and the "real causal work" (cf. Horgan 1997; Campbell 2008, 1) are found only on the microphysical level. This implies that macro-causation in a robust sense requires reductive explanations or reductions. From a metaphysical perspective, macro-causes can be taken seriously only if they turn out to be nothing but the micro-causes.

A conceptual pluralist will find such an argument not convincing. If we do not already presuppose a strong metaphysical notion of the priority of the physical,

© Springer International Publishing Switzerland 2015
D. Ludwig, *A Pluralist Theory of the Mind*, European Studies
in Philosophy of Science 2, DOI 10.1007/978-3-319-22738-2_10

there is absolutely no reason to believe that the "real causal powers" and the "real causal work" are only to be found on the (micro-)physical level. Instead, a pluralist will insist that mental (social, biological...) causes are as real as physical causes and there is no need to save mental causes through reductions.

Even if some of Kim's formulations suggest priority claims that will be rejected by conceptual pluralists, we should not be too quick in rejecting a problem of mental causation. Kim's presentation of the problem of mental causation is not limited to dubious references to "real causal powers" or "real causal work" but based on what is known as the "causal exclusion argument" (Kim 1993, 250–255, 281–292). Furthermore, it seems that this argument does not depend on priority claims that are rejected by conceptual pluralists as simple examples of the exclusion problem illustrates. Consider a headache being the cause for taking a pain killer. While the headache seems to be causally relevant, there is also a purely biological explanation of the behavior. Thus, there are at least two potential causes:

(C1) The headache is the cause for taking a pain killer
(C2) The biological process b is the cause for taking a pain killer

Reductive physicalists can offer a simple explanation of the apparent overdetermination: a headache is *nothing but* a biological process, and (C1) can be reduced to (C2). Furthermore, both (C1) and (C2) can be reduced to a microphysical cause (C3). There is no problem of overdetermination because both (C1) and (C2) are *nothing but* the fundamental microphysical cause (C3). Non-reductive theories cannot make this move and seem to be committed to real overdetermination. However, a systematic overdetermination would be utterly mysterious as it would imply that *every* mental cause is accompanied by a metaphysically distinct biological cause. And if non-reductivists want to avoid this kind of overdetermination, they seem to have to choose between the causal efficiency of the mental and the physical properties.

To see why this kind of systematic overdetermination would be mysterious, it is helpful to have a look at real world examples of overdetermination. Suppose that a house catches fire and there are two causes. On the one hand, lightning has struck, and on the other hand an arsonist has set a fire. This is a case of genuine overdetermination, which may be improbable but certainly not impossible. However, consider the following counterfactual law: *every time* a house is set on fire by a lightning strike, it is also set on fire by an arsonist. This kind of systematic overdetermination would be utterly mysterious and we would certainly demand an explanation for it.

According to Kim, non-reductive theories of the mind have to assume a systematic overdetermination that is analogous to the case of the lightning strike and arsonist. Furthermore, non-reductive theories do not have an explanation of *why* mental causes are always accompanied by biological causes. Kim argues that the only way a non-reductive theory can avoid the metaphysically implausible and bizarre idea of systematic overdetermination is to give up either (C1) or (C2). Giving up (C2) would come at the very high price of claiming that there is no sufficient biological or physical cause of mentally caused behavior. According to Kim, this rejection of the "causal closure of the physical domain" cannot be considered a serious option, which leaves non-reductive theories of the mind with the rejection of (C2). However, if non-reductive theories give up (C2), they end up with epiphenomenalism and the causal inefficacy of the mental.

Moving from a this intuitive presentation of the problem of causal exclusion to a more canonical formulation, it appears that the "causal efficacy of mental properties is inconsistent with the joint acceptance of the following four claims: (i) physical causal closure (ii) causal exclusion, (iii) mind-body supervenience, (iv) mental/physical property dualism – the claim that mental properties are irreducible to physical properties" (Kim 2005, 21–22). Assuming that a conceptual pluralist does not want to commit herself to a controversial rejection of (i) or (ii), we are left with a rejection of either causal exclusion or non-reductivism. However, systematic overdetermination would be extremely puzzling as my example of example of a different causes of the lightning strike and arsonist illustrates. Therefore, it seems that we are left with the rejection of non-reductivism and that the relaxed non-reductive attitude of conceptual pluralism is challenged by the exclusion argument.

While Kim assumes that the exclusion argument affects all varieties of non-reductivism, I think that conceptual pluralists (contrary to dualists) can respond to the argument in a straightforward way. Recall that conceptual pluralism is *not* a version of dualism or strong metaphysical pluralism (cf. Garrett 1998 who makes a related point). Different conceptual frameworks do not refer to metaphysically distinct realms of reality, but describe the same reality in terms of different but equally fundamental ontologies. And if there is no metaphysical gap between the physical and mental, then there is also no systematic overdetermination by metaphysically distinct causes.

To clarify this objection against the causal exclusion argument, I want to suggest a short thought experiment. Imagine two biologists investigating a declining hedgehog population in Malta. While the biologists conduct their field work together, they write two separate research reports. Surprisingly, they come to contradictory results:

Biologist I: Kites are the cause of the declining hedgehog population. Since kites hunt hedgehogs and are common in Malta, the hedgehog population is under pressure.

Biologist II: Hawks are the cause of the declining hedgehog population. Since hawks hunt hedgehogs and are common in Malta, the hedgehog population is under pressure.

A closer look at the research reports makes the situation even more puzzling. The biologists made most of their observations together and their descriptions are consistent with each other in almost every detail. Only one difference stands out: whenever one of the biologists reports an attack by a kite, the other biologist describes an attack by a hawk. The two research reports suggest systematic overdetermination of kite- and hawk-attacks on hedgehogs. How is that possible?

Of course, we can assume that one of the biologists is wrong and reject one of the research reports. If we accept both research reports, we seem to be confronted with a bizarre case of overdetermination. Are hawks and kites hunting together for hedgehogs? This is implausible as members of the order Falconiformes usually do not hunt in groups. Furthermore, how is it possible that one biologist only observed the hawks and the other one only observed the kites? It seems we are confronted with a situation analogous to the problem of mental causation and the competition between (C1) and (C2). Either we reject one of the causes or we find ourselves stuck with an implausible and bizarre case of overdetermination.

There is, however, a loophole. Perhaps, the biologists do not describe metaphysically distinct and competing causes for the declining hedgehog population, but rather describe causes in different ways. It is not hard to imagine how this could be the case: one biologist considers kites to belong to the genus of hawks, while the other describes kites and hawks as two different genera. Therefore, one of the biologists claims that hawks are the cause for the declining hedgehog population, while the other insists that kites and not hawks must be considered responsible. If we assume that their taxonomies are equally acceptable, we can accept that their research reports are equally correct without creating a problem of implausible overdetermination.

The obvious moral of the thought experiment is that different biological ontologies lead to different causal descriptions, but not to instances of genuine overdetermination or causal competition. Different biological ontologies do not causally compete with each other as they simply describe the causes in terms of different conceptual frameworks. If this is a case of overdetermination, it is an entirely harmless linguistic overdetermination.

Compare this biological example of linguistic overdetermination with cases of genuine overdetermination such as the house set on fire by a lightning strike and an arsonist. The crucial question is which example is the more appropriate model for mental causation. Kim's exclusion argument presupposes that non-reductive theories of mind have to model mental causation in analogy to the fire example. This is certainly a plausible claim in the case of dualism and a strong metaphysical pluralism. However, if we accept conceptual pluralism, the biological example of the declining hedgehog population is the more appropriate model. There are not two metaphysically distinct causes but two different and equally fundamental ways to describe the causes.

Analogous arguments can be made with respect to other cases of conceptual relativity. For example, consider two psychiatrists who work with the same patient during the publication of the third edition of the Diagnostic and Statistical Manual of Mental Disorders (DSM III) in 1968. One of many changes of the DSM III (cf. Wilson 1993) was the elimination of hysteria from psychiatric ontologies and its replacement with more fine-grained entities such as such as somatoform disorder or conversion disorder (Ludwig 2014). Assume that one psychiatrist still works with the DSM II and claims that the patients behavior is caused by hysteria while the other psychiatrist already works with the DSM III and claims that the patients behavior is caused by conversation disorder. Again, it seems that we have an entirely harmless form of linguistic overdetermination that does not involve metaphysically distinct entities that somehow compete for causal relevance but rather two compatible descriptions of the causes of the patient's behavior in terms of different psychiatric ontologies.

Given that I have argued for conceptual pluralism in philosophy of mind, the step from my biological and psychiatric examples to mental causation is straightforward: the irreducibility of our phenomenal ontologies does not imply that we have to assume that the phenomenal and physical are metaphysically distinct in any substantive sense that would create a problem of overdetermination. Phenomenal and physical accounts of causation do not compete with each other as they simply describe causes in terms of different ontologies. If this is a case of overdetermination, it is an entirely harmless linguistic overdetermination.[1]

References

Baumgartner, Michael. 2009. Interventionist Causal Exclusion and Non-reductive Physicalism. *International Studies in the Philosophy of Science* 23 (2): 161–78.

Campbell, John. 2007. An Interventionist Approach to Causation in Psychology. In *Causal Learning,* eds. Alison Gopnik and Laura Elizabeth Schulz, 58–66. Oxford: Oxford University Press.

Campbell, Neil. *Mental causation: A nonreductive approach.* Frankfurt: Peter Lang, 2008.

Eronen, Markus I. 2011. *Reduction in Philosophy of Mind: A Pluralistic Account.* Berlin: Walter de Gruyter.

Garrett, Brian Jonathan. 1998. Pluralism, Causation and Overdetermination. *Synthese* 116 (3): 355–378.

Glennan, Stuart. 2010. Mechanisms, causes, and the layered model of the world. *Philosophy and Phenomenological Research* 81 (2): 362–381.

Horgan, Terry. 1997. Kim on mental causation and causal exclusion. *Noûs* 31 (11): 164–184.

Kim, Jaegwon. 1993. *Supervenience and Mind: Selected Philosophical Essays.* Cambridge: Cambridge University Press.

Kim, Jaegwon. 2005. *Physicalism, or Something Near Enough.* Princeton: Princeton University Press.

Ludwig, David. 2014. Hysteria, Race, and Phlogiston. A Model of Ontological Elimination in the Human Sciences. *Studies in History and Philosophy of Science Part C: Studies in History and Philosophy of Biological and Biomedical Sciences* 45: 67–77

Pereboom, Derk. 2002. Robust nonreductive materialism. *The Journal of Philosophy* 99 (10): 499–531.

Raatikainen, Panu. 2010. Causation, Exclusion, and the Special Sciences. *Erkenntnis* 73 (3): 349–63.

Wilson, Mitchell. 1993. DSM-III and the Transformation of American Psychiatry: a History. *American Journal of Psychiatry* 150: 399–399.

Woodward, James. 2008. Mental Causation and Neural Mechanisms. In *Being Reduced,* eds. Jakob Hohwy and Jesper Kallestrup, 52–74. Oxford: Oxford University Press.

---

[1] Of course, this response does not provide a positive account of non-physical causation and may be combined with more specific proposals such as an interventionist framework. Indeed, interventionist responses to the problem of mental causation have become increasingly popular in recent years (e.g. Campbell 2007; Woodward 2008; Raatikainen 2010) and may be seen as natural allies of conceptual pluralism due to their permissive implications regarding non-physical causes (cf. Eronen 2011). At the same time, I think that it is helpful to distinguish my pluralist response from interventionist accounts as my response is not affected by some more recent objections against interventionist accounts of mental causation (e.g. Baumgartner 2009) and is also available to philosophers who find interventionism implausible.

Chapter 11
Epilogue Metaphysics in a Complex World

My case for a pluralist theory of mind has been based on general claims about the diversity of scientific practice. Although ontological and epistemic unifications plays an important role in many research contexts, we have no good reason to assume that they have to be successful everywhere. Instead, scientific explanations range from traditional forms of theory reductions and reductive explanations to clearly non-reductive forms of integration of scientific methods, models, and ontologies.

Large parts of this book have been concerned with challenges of global unification efforts and its underlying metaphysical motivations such as the ideal of exactly one fundamental ontology. One may therefore worry that a thoroughly pluralist proposal does not leave any room for substantive issues in philosophy of mind and metaphysics. As soon as we have given up traditional placement problems and the ideal of one fundamental ontology, there are no interesting metaphysical issues left. I do not share this anti-metaphysical sentiment and want to conclude by arguing that pluralist metaphysics can be highly productive in engaging with the diversity of explanations and ontologies in scientific practice.

11.1 Towards an Empirically Grounded Philosophy of Mind

Cognitive science provides one of the most vivid examples of the diversity of explanations that we find in scientific practice. One obvious starting point for a discussion of this diversity is the overwhelming number of fields and subdisciplines in cognitive science such as artificial intelligence, behavioral genetics, biological psychology, clinical neuropsychology, computational neuroscience, cognitive neuroscience, cognitive psychology, cybernetics, molecular neuroscience, neurolinguistics, neuropharmacology, psycholinguistics, psychometrics, psychophysics, robotics, social neuroscience, and so on. Issues of unification and integration do not only arise if we consider the relation between these fields. Instead, many of them are

© Springer International Publishing Switzerland 2015
D. Ludwig, *A Pluralist Theory of the Mind*, European Studies
in Philosophy of Science 2, DOI 10.1007/978-3-319-22738-2_11

already the result of integration efforts that are concerned with relations between entities that belong to different levels of organization (e.g. Abney et al. 2014; Looren de Jong 2002; Craver 2007).

This diversity of explanatory strategies in cognitive science suggests an empirically grounded philosophy of mind that has little in common with global reductionist or anti-reductionist approaches. If we employ strong notions of reduction and reductive explanation that aim at ontological and epistemic unification, it is reasonable to assume that we will at least occasionally encounter substantive forms of irreducibility. For example, I have argued that many biological kinds such as ecological species have little to do with microphysical kinds and therefore make ontological and epistemic unification implausible. In cognitive science we occasionally encounter similar situations (cf. Sect. 4.2). For example, it is very unlikely that every useful psychiatric kind corresponds with a coextensive neural kind and a general rejection of multiple realization in psychiatry would have to employ a very restrictive notion of natural kinds that greatly reduces the diversity of psychiatric kinds. Furthermore, externalist research programs in cognitive science are mostly concerned with patterns that do not neatly match biological kinds and set limits for unification between cognitive and biological levels of organization.

However, this does not mean that cognitive science is fundamentally disunified and that there is nothing interesting to be said about the relation between entities of different cognitive and biological levels of organization. First, reductive explanations can play an important role in cognitive science (Sects. 6.5 and 6.6), even if they do not lead to global ontological and epistemic unification (Sect. 6.7). Second, debates about issues such as mechanistic explanations illustrate the diversity of meaningful relations between different areas of cognitive science. Many of these explanations do not fit the classical reducibility vs. irreducibility dichotomy but seem to require a gradual understanding in which traditional accounts of reducibility and irreducibility are mostly idealized ends of a gradual scale.

For example, Bechtel presents his account of mechanistic explanation as "reductionistic insofar as it appeals to the components of a mechanism to explain its activity." At the same time, he stresses that "insofar as the phenomenon generated by a mechanism depends upon the organization of the parts and the conditions impinging upon the mechanism from without, mechanistic explanation also recognizes the autonomy of higher-level investigations." (2006, 192). Craver (2005, 2007), however, presents mechanistic explanation as an alternative to reductionism by pointing out the multilevel character of mechanisms. For example, in his discussion of Long-Term Potentiation (or LTP) in memory research, Craver stresses the integration of different levels of analysis that incorporate bottom-up and top-down approaches: "This mechanistic shift involved coming to see LTP not as identical to memory or as kind of memory, but rather as a component in a multilevel memory mechanism" (2006, 383).

As suggested by my discussion in Sect. 6.3, the question whether mechanistic explanation is *really* reductionistic is not all too helpful. While there is obviously room for substantive disagreement about the structure of mechanistic explanations, the vagueness of the term "reduction" also allows different rhetorical strategies. Godfrey-Smith provides a nice example of this:

> I should note that, once this picture is in place, the fate of the term "reduction" can become unclear. At a 2006 symposium on the relation between philosophy of science and philosophy of mind at Boston University, Steven Horst and William Bechtel gave talks that, on these points at least, presented fairly similar pictures of how the relevant areas of science operate, and the deficiencies of more traditional views. But Horst saw his message as anti-reductionist; his talk was titled "Beyond Reduction". Bechtel, in contrast, saw himself as describing what real reductionist work, as opposed to the philosophers' image of it, is like. In discussion, Bechtel (and Paul Churchland) argued that the term "reduction" is entirely natural for this kind of scientific work. (Godfrey-Smith 2008, 57)

Even if the diversity of scientific explanations often challenges a clear dichotomy between reducibility and irreducibility, there are also research programs in cognitive science that self-identify as "reductionistic". Bickle's (2003, 2006) discussion of "ruthless reductions" provides a helpful example as he distances himself from philosophical models of reduction but still stresses the importance of what he calls "ruthless reductions" in molecular neuroscience. Based on a methodology that is strikingly similar to what I have described as a "naturalism of scientific practice", Bickle urges us to ignore large parts of the philosophical debate about reduction and to look at clear examples of reductionistic inquiries in scientific practice.

Bickle identifies molecular neuroscience as a case study of a reductionist field in cognitive neuroscience and develops his account of "ruthless reductions" on the basis of current research on mind-molecule links. Of course, one can challenge Bickle's characterization of molecular neuroscience as ruthlessly reductionistic (cf. Looren de Jong 2006). For example, Craver's discussion of the molecular mechanisms of LTP challenges Bickle's account of neuroscience as a ruthlessly reductive bottom-up enterprise. Instead, Craver suggests that we need to consider both bottom-up and top-down approaches in order to gain a full understanding of the multi-level character of research on LTP and memory. He therefore concludes that "the history of the LTP research program provides a clear historical counter example to those (such as Oppenheim and Putnam 1958; or Bickle 2003) who tout reduction as an empirical hypothesis" (2006, 384).

However, there is also another difficulty in the step from ruthless reductions to ruthless reductionism. Even if we accept Bickle's case studies as examples of ruthless reductions, it is still far from clear why we should embrace a more general ruthless reduction*ism*. Arguably, the label "ruthless reductionism" is ambiguous in allowing a weak and a philosophically ambitious interpretation. The weak interpretation considers molecular neuroscience as an example of ruthless reductionism simply because ruthless reductions are common in this field. This weak interpretation does not imply that ruthless reductionism is a helpful model for neuroscience, cognitive science, or even science in general. Instead, it suggests that we should consider the scope of ruthless reductions an empirical question and it leaves room for areas in cognitive science that rely on different forms of scientific explanation.

Furthermore, it seems easy to find examples in cognitive science that have little use for ruthless reductions. Consider, for example, my discussion of research on problem solving in the context of externalist accounts of cognition. Many cognitive scientists are drawn to externalist accounts because they are concerned with interesting patterns in areas such as information retrieval and problem solving that incorporate both biological and non-biological information storages. Arguably, these patterns will not reduce to a molecular level and therefore constitute one limit of ruthless reductionism in cognitive science. On a more general level, the argument from horizontal pluralism suggests a liberal account of psychological ontologies that neither supports a general reductionism nor anti-reductionism. While Bickle's detailed discussions of "mind-molecule links" suggest that ruthless reductions are attractive with respect to some psychological kinds, there is very little reason to believe that *every* legitimate psychological kind will neatly correspond with broadly coextensive molecular kinds.

Of course, none of this contradicts a weak interpretation of "ruthless reductionism" that simply states the importance ruthless reductions in some areas of scientific practice. The situation is obviously different with a philosophically ambitious ruthless reductionism that aims at a general model of reduction in neuroscience, cognitive science, or even a global model of reduction in science. Such an ambitious variant of "ruthless reductionism", however, will have to challenge entities such species (Sect. 4.1), psychiatric kinds such as delusion (Sect. 6.4), or extended memory (Sect. 4.2) that seem to lack appropriate "molecule links". Although an ambitious ruthless reductionist could endorse an eliminative strategy regarding alleged counter-examples and argue that these entities do not *really* exist, I find it difficult to see how such a strategy could be successful in Bickle's own "metascientific" framework.

For example, cognitive scientists often use externalist accounts of memory because they are interested in patterns that involve not just the brain but also include non-biological information storages. Any justification of the claim that these patterns do not really exist would have to take a huge step from a metascientific description to an ambitious metaphysical evaluation of the status of scientific entities. My discussion of the metaphysical priority of the physical illustrates how such an argument could look like. For example, one could insist that an austere physicalist ontology requires that we eliminate all entities and patterns that are not captured by ruthless reductions.

However, Bickle does not want to be metaphysician and rather insists on the authority of scientific practice. Given this methodological framework, it is very hard to see how one could successfully argue for an eliminative stance towards the entities of successful empirical sciences. In other words: even if we accept that ruthless reductions are an important form of explanation in cognitive science, there is little reason to accept a global ruthless reductionism that would be incompatible with a substantive pluralism.[1]

---

[1] Despite Bickle's rhetoric of "ruthlessness", his discussions of a metascientific methodology actually indicates a rather weak interpretation of "reductionism". Bickle does not only avoid specific ontological commitments but declares: "Trust scientific practice and let ontological chips fall where they may" (2003, 32). Furthermore, his metascientific methodology comes even closer to

The diversity of scientific explanations in cognitive science provides a helpful starting point for the ideal of an empirically grounded philosophy of mind. My discussion has been highly critical of large parts of philosophy of mind and its obsession with placement problems such as phenomenal consciousness, intentionality, or self-knowledge. I have argued that this focus on placement problems presupposes a dubious metaphysical picture that is simply not supported by our best understanding of science. An empirically grounded philosophy of mind strongly differs from this traditional approach that has dominated philosophy of mind since the second half of the twentieth century. More specifically, it turns the relation between science and metaphysics upside down. Instead of starting with a metaphysical picture and the question how some idealized science can validate it, an empirically grounded philosophy of mind assumes that metaphysics should be shaped by engagement with scientific practice.

While philosophical debates about the mind-body problem usually continue with a traditional approach that starts out with metaphysics instead of scientific practice, current debates in philosophy of neuroscience and cognitive science are increasingly based on a different methodology. A helpful example are debates about reduction and the growing discontent with philosophical models of theory reductions and reductive explanation that are shaped by metaphysical requirements instead of actual scientific explanation. Bickle (2006, 34) nicely illustrates this attitude by stressing the importance of "analysis of real reductionism in real scientific practice, as contrasted with artificial accounts of scientific reductionism that rest instead on philosophical assumptions about 'what reduction has to be'. So characterized, that project strikes me as inherently reasonable. So why is it virtually non-existent in contemporary philosophy?"

While Bickle's complaint is justified in the context of contemporary philosophy of mind, there is also a growing number of philosophers who stress the importance of an empirically grounded account of explanation in cognitive science as the current wave of interest in mechanistic explanations shows. Furthermore, the differences between philosophers like Bechtel, Bickle, and Craver illustrate that there is

---

my discussion of conceptual relativity by endorsing Carnap's (1950) internal-external distinction and rejecting "'external' questions about the existence and nature of 'theory-independent ontology.' Rather, because a reasonable explanatory goal is to understand practices 'internal' to important current scientific endeavors and the scope of their potential application and development" (2006, 32). Given these striking similarities between Bickle's metascientific methodology and my own presentation of a naturalism of scientific practice, one may suspect that our main difference is merely rhetorical. While Bickle focuses on the reductionistic parts of science, I stress the diversity of reductionistic and non-reductionistic projects in a more general perspective on scientific practice. However, our differences are not *only* rhetorical but arguably extend to the question of how reductionistic contemporary science is. While Bickle would probably object that I do not take the importance of ruthless reductionism in scientific practice seriously enough, it seems to me that Bickle's discussion often suffers from a confirmation bias: he finds scientific practice to be ruthlessly reductionistic because he is especially interested in the ruthlessly reductionistic parts of science. However, a thoroughly metascientific approach would have to avoid biased samples by letting scientific practice determine the importance of reductionistic and non-reductionistic research programs.

room for substantial disagreement in an empirically grounded philosophy of mind.
Even if we agree that metaphysics should be shaped by scientific practice, there is
obviously still a lot of room for disagreement about the philosophical lessons of
engagement with scientific practice that reach from a liberal pluralism to a "ruthless
reductionism".

Finally, even if we assume that an empirically grounded philosophy of mind will
be thoroughly pluralistic, there still remain important philosophical issues. A reasonable pluralism will not be limited to claims of disunity and irreducibility but
acknowledge the importance of both reductive and non-reductive forms of explanation. Even if we reject the idea that every legitimate entity has to be explained in
terms of one fundamental ontology, there can be still not doubt that there are meaningful forms of epistemic and ontological unification and integration. A substantive
pluralist philosophy of mind will engage with both the scope and the limits of these
diverse forms of scientific explanation.

11.2   Metaphysics and Unification

I have argued that conceptual pluralism challenges placement problems in philosophy of mind but leaves room for an empirically grounded philosophy of mind that
is concerned with the diverse relations between ontologies and methodologies in the
cognitive sciences. The same considerations apply to metaphysics in general.
Indeed, my proposal of conceptual pluralism is incompatible with a certain type of
metaphysics that starts with the assumption of a (e.g. physicalist, dualist, idealist)
base of fundamental entities and is mostly concerned with placement problems of
fitting non-fundamental entities into this base.

However, pluralists do not need to accept an identification of metaphysics with
this problematic ideal of exactly one fundamental account of reality. Instead, they
can insist that they are also proposing a metaphysical picture, even if this picture
does not entail the ideal of exactly one fundamental and unified account of reality.
In analogy to Putnam's (2004, 21) distinction between ubiquitous ontological issues
and Ontology with a capital "O", one may also distinguish between perfectly legitimate metaphysical projects that are concerned with the structure of reality and a
Metaphysics with a capital "M" that insists on exactly one fundamental account of
the structure of reality.

While conceptual pluralists can insist that a satisfying metaphysical account will
include diverse forms unification, integration, and disunity, one may wonder
whether such a pluralism actually leaves a room for substantive metaphysical
inquiry. For example, Ladyman and Ross (2007) argue for an "understanding of
metaphysics as consisting in unification of science" (2007, 5). Pluralist positions
that stress the importance of integration without unification or even stronger variants disunity therefore seem to have the unfortunate consequence of undermining
the very possibility of metaphysics.

I have to admit that I find Ladyman and Ross' definition of metaphysics in terms of unification puzzling. It is clearly true that many metaphysical projects aim for unification in one way or another but why should we exclude disunity from metaphysics altogether? This question is especially pressing in the context of Ladyman and Ross' own naturalistic approach that is sharply contrasted with large parts of contemporary analytic metaphysics. Given Ladyman and Ross' naturalistic methodology, metaphysical accounts of unification have to be developed on the basis of engagement with unification in the empirical sciences. While Ladyman and Ross correctly stress that "the unification of a range of phenomena within a single explanatory scheme is often a goal of scientific theorizing" (2007, 71), no reasonable pluralist would reject this claim. Instead, pluralists will argue that unification is important but not equally important in all areas of scientific practice. Indeed, there are also many research areas that are dominated by weaker forms of integration or "interfield-theories" (Darden and Maull 1977). Furthermore, we may also encounter stronger forms of disunity when considering the relation between entities in largely unrelated fields such as ecology and particle physics.[2]

Even if a definition of metaphysics in terms of unification is unsatisfying, Ladyman and Ross' discussion entails important observations about the relation between science and metaphysics. While their naturalist methodology implies that metaphysics should be shaped by empirical research, metaphysics must in some way go beyond paraphrasing the results of science. If metaphysicians would not have anything interesting to say that has not already been spelled out by empirical scientists, there would be no room for metaphysics. If I understand Ladyman and Ross correctly, unification has a defining role in their account of metaphysics *because* a fundamentally disunified universe would leave nothing interesting to say beyond the results of the empirical sciences: "if the world were fundamentally disunified, then discovery of this would be tantamount to discovering that there is no metaphysical work to be done: objective inquiry would start and stop with the separate investigations of the mutually unconnected special sciences. By 'fundamentally disunified' we refer to the idea that there is no overarching understanding of the world to be had" (2007, 6).

It is indeed helpful to conceive metaphysics as a synthetic discipline that aims for an "overarching understanding of the world" (or parts of the world) on the basis of our often staggeringly specialized empirical knowledge about reality. However, this kind of "overarching understanding" does not need to take the form of global

---

[2]One may object that unification is of crucial importance in fundamental physics and is limited only in less fundamental areas of the "special sciences". Indeed, Ladyman and Ross do not assign equal epistemic authority to all areas of scientific practice but introduce a *Primacy of Physics Constraint* (PPC) as a "methodological rule" according to which "practitioners of special sciences at any time are discouraged from suggesting generalizations or causal relationships that violate the broad consensus in physics at that time, while physicists need not worry reciprocally about coherence with the state of the special sciences" (2007, 38). Even under the assumption of PPC, however, it is not clear why limits of unification in the "special sciences" should not be taken seriously. Certainly, limits of unification in the integration of biological theories do not violate the "broad consensus in physics" in the sense of PPC.

unification. Our accounts of the world could be partly unified and partly disunified in the sense that we encounter all kinds of interesting relations that sometimes lead substantive unification but sometimes also to other forms integration or disunity. Investigating both the scope and the limits of unification would certainly add to our overarching understanding of the world and should therefore count as proper metaphysics.[3]

Even if Ladyman and Ross define metaphysics too narrowly as unification, their more general idea of metaphysics as aiming at an "overarching understanding" is helpful in explaining the relation between science and metaphysics. A pluralist metaphysics is necessary if and only if the world turns out to be considerably more complex than imagined by metaphysicians who aim for global unification. However, a reasonable pluralism will not be limited to claims of disunity and criticism of reductionism. Instead, it will aim at providing an overarching understanding by developing a systematic account of the numerous forms of unification, integration, and disunity. Even if our best scientific knowledge suggests that the world is considerably more complex than assumed by proponents of global unification, we should still aim for synthetic approach that moves beyond simply listing unrelated scientific theories and entities. Metaphysical orientation may be more difficult than suggested by the grand metaphysical narratives of physicalism, dualism, and idealism but that does not mean that metaphysical orientation is impossible.

A pluralist theory of the mind exemplifies this ideal of pluralist metaphysics. Far from *only* insisting on irreducibility and ontological disunity, a comprehensive pluralism in philosophy of mind would provide an empirically grounded account of the diverse relations between our scientific and non-scientific forms of knowledge about the mind. These relations may include traditional or "ruthless" reductions, reductive explanations without reductions, different varieties of mechanistic explanations, non-mechanistic causal explanations, and stronger forms of irreducibility and disunity. Clearly, this book does not provide such a comprehensive account but has been mostly an exercise in *meta*philosophy of mind. Instead of providing a detailed account of the relations between different kinds of knowledge about mind and cognition, I have been mostly concerned with the general aims and methods of philosophy of mind. Accepting the proposed pluralist framework will therefore neither undermine philosophy of mind nor answer all philosophically interesting

---

[3] Arguably, Ladyman and Ross are still right that a "fundamentally disunified universe" that does not allow *any* interesting forms of unification and integration would undermine the very possibility of metaphysics. However, this is a caricature not only of mainstream "integrative" pluralism (e.g. Mitchell 2003) but also of pluralists who emphasize disunity. Ladyman and Ross therefore seem to challenge a straw-man who does not only reject global unification ideals in science and metaphysics but replaces global unification with global disunity. For example, Ladyman and Ross seem to identify Dupré (1993) as a pluralist who undermines the possibility of metaphysics by insisting on a "fundamentally disunified universe". However, Dupré has clarified on numerous occasions that this is misunderstanding of his arguments that are directed against a global ontological unification but do not deny the importance of reduction or even weaker forms of unification in science. For example, in an exchange with Dennett, Dupré stresses "I am second to no one in my enthusiasm for even some very reductionist parts of science. Our disagreement is, in the end, just on which parts are good and which parts are bad" (2005, 695).

questions about the mind. Instead, it shifts the focus from placement problems to an empirically grounded philosophy of mind that is concerned with our diverse empirical knowledge about the mind and its implications for our self-understanding.

References

Abney, Drew, Rick Dale, Jeff Yoshimi, Chris Kello, Kristian Tylén, and Riccardo Fusaroli. 2014. Joint Perceptual Decision-Making: A Case Study in Explanatory Pluralism. *Theoretical and Philosophical Psychology* 5 (330): 1–19.

Bechtel, William. 2006. Reducing psychology while maintaining its autonomy via mechanistic explanation. In *The matter of the mind: Philosophical essays on psychology, neuroscience and reduction* eds. Maurice Schouten and Huib Looren de Jong, 172–198, Chichester: Wiley.

Bickle, John. 2003. Philosophy and Neuroscience: *A Ruthlessly Reductive Account*. Boston: Kluwer.

Bickle, John. 2006. Ruthless Reductionism in Recent Neuroscience. *Systems, Man, and Cybernetics, Part C: Applications and Reviews, IEEE Transactions On* 36 (2): 134–40.

Carnap, Rudolf. 1950. Empiricism, Semantics, and Ontology. *Revue Internationale de Philosophie* 4: 20–40.

Craver, Carl F. 2005. Beyond reduction: Mechanisms, multifield integration and the unity of neuroscience. *Studies in History and Philosophy of Science Part C: Studies in History and Philosophy of Biological and Biomedical Sciences* 36(2): 373–395.

Craver, Carl F. 2007. *Explaining the Brain*. Oxford: Oxford University Press.

Darden, Lindley, and Nancy Maull. 1977. Interfield Theories. *Philosophy of Science* 83: 43–64.

Dupré, John. 1993. *The Disorder of Things: Metaphysical Foundations of the Disunity of Science*. Cambridge, Mass.: Harvard University Press.

Dupré, John. 2005. You Must Have Thought This Book Was About You: Reply to Daniel Dennett. *Philosophy and Phenomenological Research* 70(3): 691–695.

Godfrey-Smith, Peter 2008. Reduction in Real Life. In *Being Reduced*, eds. Jakob Hohwy and Jesper Kallestrup, 52–74, Oxford: Oxford University Press.

Ladyman, James, and Don Ross. 2007. *Every Thing Must Go: Metaphysics Naturalized*. Oxford: Oxford University Press.

Looren de Jong, Huib. 2002. Levels of Explanation in Biological Psychology. *Philosophical Psychology*, 15 (4): 441–462.

Looren de Jong, Huib. 2006. Explicating Pluralism: Where the Mind to Molecule Pathway gets off the Track – Reply to Bickle. *Synthese* 15: 435–443.

Mitchell, Sandra D. 2003. *Biological Complexity and Integrative Pluralism*. Cambridge: Cambridge University Press.

Oppenheim, Paul, and Hilary Putnam. 1958. Unity of science as a working hypothesis. *Minnesota Studies in the Philosophy of Science* 2: 3–26.

Putnam, Hilary. 2004. *Ethics Without Ontology*. Harvard: Harvard University Press

Printed in the United States
By Bookmasters